GRAND SLAM

An international bridge tournament
between the United Kingdom and
the United States

Contents

Introduction

THIS IS the story of a Bridge match between the United Kingdom and the United States recorded by the BBC at a country house in Gloucestershire. It was the ideal setting for a game invented only 60 years ago by the American millionaire Harold S. Vanderbilt and first played on his luxury yacht by players presumably bored with the imprecision of the older form of Auction Bridge. But, contrary to myth, Bridge is not the preserve of the rich; a somewhat indigestion-provoking accompaniment to tea and crumpets in front of the drawing room fire. It has been played everywhere, from palaces to prisoner-of-war camps because, quite simply, it is the best of all card games.

It is not a game of super-science, and neither can it be played only by accomplished mathematicians; and if that seems unbelievable, watch four good players trying to divide a restaurant bill equally. Anyone who can count up to 13 without the aid of a calculator can play it.

Unfortunately, like so many other sports and games, its clarity has been muddied by misguided and ever-theoretical experts; it may not be quite as easy as 'Snap' but it is no more difficult that any other game in which a modicum of common sense and concentration is needed. It is, however, a game which excites considerable passion; there is one case recorded of a player who shot her husband after yet another ill-judged play – and was freed by an American judge, obviously a Bridge player himself, on the grounds of justifiable homicide.

One of the greatest charms of the game is that no hand lasts very long, and if you make the most frightful mistake, you have only to live with it for the few minutes until the next deal, unlike the unfortunate chess player who can blunder at move three and suffer the consequences for the next two and a half hours.

Above all, it is a partnership game, and to succeed one has to be if not a little in love with, at least respectful and understanding of one's partner. The game bristles with pitfalls and traps and everybody falls into them, even players of the calibre of those taking part in the match that follows. Nobody makes mistakes deliberately, but unless you are playing for stakes higher than you can afford, most blunders are funny rather than disastrous.

There was one change in the British team that won the first televised Grand Slam Trophy Match; the vulpine Jeremy Flint, having been invited to commentate, was replaced by another international, Tony Priday.

Priday would make most Americans, and many of the British for that matter, feel instantly one down. He is tall, grey-haired, distinguished and impeccably dressed – the sort of man who would wear a tie at breakfast even when he was on his own. He is also amusing, polite and might appear ripe to be mugged at the Bridge table, but anyone who tried would be making a great error. Beneath the velvet lurks a mind of iron and if his critics deride him for being, on occasion, over-cautious he will politely agree and diffidently refrain from pointing out that his record in international Bridge is second to very few in Britain.

His partner, Claude Rodrigue, is stocky and pugnacious and not a man to be trifled with under any circumstances. He has honed his aggression and technique by playing high-stake Rubber Bridge where he is renowned for suffering fools gladly only when they are his opponents. He and Priday are rather like expert inquisitors; Priday lulls with silken-tongue and steely-eyed charm while Rodrigue hovers with a sock full of wet sand waiting to club any indiscretion.

The British women, Nicola Gardener and Pat Davies, have an indefinable air of defencelessness away from the table; the sort of players one feels it might be a pleasure to cut against in any club. So they are, if you are astute enough to learn from them and positively enjoy losing. Nicola Gardener is one of the finest women players in the world. Not only is she an excellent technician – truly, the daughter of her father Nico Gardener – but is equally aggressive and imaginative both in bidding and defence. Pat Davies is more studious, as becomes a lecturer in statistics, and is coolly logical and analytical about what she does even if, endearingly, she is rather more prone to error than her partner. She and Nicola are two members of Britain's team of Women World Champions (Port Chester, N.Y. State, 1981), and there are many discerning commentators who believe Pat Davies is still improving.

They succeeded to their title from the American women, Jackie Mitchell and Gail Moss, who would dearly like it back. Jackie Mitchell seems to live on an apple and a five-mile jog a day and is as utterly self disciplined at the table as she is away from it. She used to give the impression of being a bundle of nerves, terrified of the criticism of her husband Victor (an outstanding American international) and many other things besides. She passes much of her time at the table doing embroidery as though determined to remain detached, but those who trifle with her do so at their peril. If her bidding sometimes appears orthodox, or a little too conservative, her card play and defence are of the very highest quality.

Gail Moss, her partner, sometimes strays from orthodox. Of all the four women players, she was described by one commentator as being

7

the most feminine, 'and I don't mean that as a compliment,' he added sourly. 'Flair, or women's instinct, or call it what you will, usually means no more than simply flouting some well-established principle – at least at the Bridge table. If it works, the feminist points to "flair". If it doesn't, men are usually too chivalrous to class it as downright ignorance.' Male chauvinism is alive and well and living in most Bridge clubs! Gail Moss is decidedly lively in her defensive play, always looking for the chance to attack, but there is the impression – reinforced by some of her thought tracks – that Jackie Mitchell has tempered her partner's exuberance with some sharp reproofs.

In the same way as there is a mould for American golfers – they all seem to be six foot six, blond, hit the ball 350 yards and automatically sink 75 foot putts – so there is a mould for American bridge players. Neil Silverman is cast in that mould, a straightforward scientific player with excellent technique, flashes of brilliance and the ability to throw the occasional curve ball to deadly effect. He made a lot of money out of horse racing, a fortune out of the New York Stock Exchange, and many friends in Gloucestershire by playing rubber bridge with members of the BBC team without looking at his cards until the bidding was completed. Had his partners been slightly more logical, his bidding would have been impeccable. He has a brisk, no-nonsense approach to the game. Playing with Matthew Granovetter as his partner, speed is not only desirable, it is essential.

Like Silverman, Granovetter comes from New York and is perhaps the most intriguing of all the eight players taking part. He makes his living out of bridge, not teaching as do Nicola Gardener and the two American women, or as professionals might through playing Rubber Bridge with those rich enough to be able to afford to lose to them, but by hiring himself out to those prepared to pay highly to have him as a partner or as a member of their team.

He is a composer and lyricist and he plays Bridge like an artist. One moment he is suffused with extrovert optimism, the next he is submerged in gloomy introspection which leads to some unsound overbidding and some extreme conservatism. Very often, to achieve the artistically perfect result, he plays so slowly that the whole table seems frozen in some timeless still-life, but he demonstrated time and again that he was one of the best card players on either side.

A match played over 78 boards is not an isolated series of technical problems for the players. While Boards One to Six are being played in Room One, Boards Seven to Twelve are being played in Room Two. The Boards are then changed, so that a session consists of twelve boards. During that time, the players do not know what their team-mates have been doing in the other room. They know what contracts were reached in their own room, they know what the optimum contracts were, they know what brilliances and disasters they have perpetrated, but the rest is guesswork or even a mystery. The match thus becomes a test of nerve, character and stamina as well as a trial of pure technical ability. Every twelve boards, the teams went into a huddle to analyse what had happened and to discuss their tactics. Whereas their thoughts on bidding and play are public, those discussions remain largely private.

You may well feel that Bridge at the level of this match is like a gymnast's display; to be enjoyed and admired but not understood. As you read the players' own thoughts, you will be surprised, and perhaps consoled, by how often they admit that they are not really sure what is going on. Gradually the light dawns that even on this plane, Bridge is not a super science but much more a matter of logic and deduction.

The experts are doing what any novice tries to do: find out how many tricks they can make, and decide which suit, if any, to choose as trumps, by means of the code known as 'bidding.' Their codes might be more complicated than those of the average player, but they are based on the same principles. Of course, all systems are governed by the fact that the suits have an order of priority and each bid must be at a level higher than the one preceding it.

Innumerable books have been written about bidding systems, but what must have surprised spectators at this match is the number of times players asked the non-bidding opponent what is understood by their partner's bid. This does not display ignorance; it is perfectly legitimate because it is as much against the rules to confuse opponents with utterly secret, private signals as it is to kick your partner under the table or hold the cards in a particular way to show you have a specific suit.

What distinguishes an expert is his greater powers of logic and deduction and his ability to apply simple principles to apparently sophisticated situations. Take, for instance, the question of winning and losing tricks. A player who has contracted to make nine tricks, for example, has also by implication contracted to lose not more than four tricks. Every palooka knows that if he holds six cards in a suit headed by KQJ with two outside Aces, if he persists with his long suit, then after the opponents have finally taken their Ace he will make five tricks in that suit and the two Aces.

Experts apply that principle all the time; it may be deeply disguised, but the principle of losing tricks advantageously is one that underlies much of the skill and charm of the game.

The next section, 'Beginners, Please', is for those who are not quite sure whether or not a bid of 1♠ over 1♣ is legal. It also includes a glossary of conventions, an awareness if not a profound understanding of which will add to the enjoyment of 'The Match' itself.

PART ONE

Beginners, Please

CONTRACT BRIDGE is a game of war concerning four players split into two partnerships. For ease of recognition they are identified by the cardinal points of the compass, North (N), East (E), South (S) and West (W). North-South form one partnership and East-West their adversaries. Each deal is similar to a hand of whist and is a separate battle forming part of a major encounter called a rubber – hence the name Rubber Bridge. As the 52 cards in a pack are distributed 13 to each player, one at a time clockwise, it is a matter of chance whether a partnership receives a good combined holding or a bad one. If the latter, they stand to lose the battle, but not necessarily the war.

In competition bridge, luck has to be eliminated. This is achieved by fighting each battle twice, which gives rise to the name Duplicate Bridge. In this type of event four players form one team, two of them occupying the North-South positions in one room and their allies occupying the East-West positions in a separate room. Their opponents, similarly, are split North-South and East-West between the two rooms. The cards are dealt in one room and the battle is fought out but the four separate hands are not destroyed. They remain intact in their respective North, East, South and West positions. At the end of the battle each hand is placed in a board which has four compartments and is transported to the other room. There it is played again under exactly the same conditions, but this time the North-South players will be on the opposing side to the North-South players in the first room. After the second battle, the two results are compared and the difference, calculated in arithmetical terms, determines whether a team has won, lost or tied the board.

In the Grand Slam television series four players represent the United States of America and four represent the United Kingdom. A total of 78 boards are played and each week the scores on six boards are reviewed. Particular emphasis is placed in the TV programme on the more noteworthy hands, three of which are usually described in detail each week. The scores on the remaining three boards are announced, so that each week, for a period of thirteen weeks, viewers can learn what is the running score, or how the war game is progressing. At the end of the period after the play on board 78 has been described, the final score is known and the destination of the £8,000 prize money determined.

The rudiments of Contract Bridge

There are 52 cards in a pack, split into four suits – Spades (♠), Hearts (♡), Diamonds (◇) and Clubs (♣). Each suit of thirteen cards is identical, with cards ranking from Ace highest, through King, Queen, Jack, 10, 9, 8, 7, 6, 5, 4, 3, and 2, the lowest. The suits also have a rank: Spades (highest), through Hearts, Diamonds, and Clubs, the lowest. In rubber bridge each of the four players wishing to participate draws a card from the pack. The rank of these cards determines which pair plays against which other. The player drawing the highest card chooses his seat, let us say North, and his partner who selected the next higher card occupies the South seat. The pack is then shuffled by East and cut by West. North completes the cut and deals the first hand, one card at a time clock-wise, starting with East. While North is dealing, South shuffles a second pack of cards with different coloured backs from the first. These are then placed to his right, adjacent to East, denoting who is to deal on the next round.

When the deal is completed, each player picks up his thirteen cards and sorts them into suits. The dealer then has the first call, which is his initial attempt to determine the trump suit. Normally he does this by calling one of his longest suit, e.g. One Heart which means that he undertakes or *contracts* to make seven tricks (one over the halfway point – thirteen cards in each hand means thirteen tricks, but more of that later). East then has his chance in the auction and provided that he makes a higher call he can try to select the trump suit. One Spade as an overcall would be sufficient in the above example but if East's best suit is Clubs or Diamonds he must call at least Two Clubs or Two Diamonds to outrank North's opening bid of One Heart. If he does not wish to enter the bidding he must say 'Pass' or 'No Bid'. The auction proceeds clockwise round the table and keeps progressing as each player in turn outbids his opponents or even his partner. The auction ceases only when there are three Passes in rotation. At this point the play stage is entered.

Say the final contract is Four Spades to be played by South, not necessarily because he called Four Spades – his partner might have done so but because he bid Spades before his partner. This means that South will be *Declarer*, but he will be supported in his endeavour by the North hand. North, the player, however, does not participate in this battle. He is known as *Dummy* and after West makes the initial lead, he merely displays his hand (The Dummy) on the table with the trump suit nearest the West player.

11

South is the general in this particular battle. He has the responsibility of assessing his army's chances of winning the skirmish, and playing accordingly. He may play high cards and capture the enemy high cards, or he may sacrifice some of his troops in order to stage a counter-attack later. His primary objective is to make ten tricks with Spades as the trump suit.

Trick-taking

Each trick consists of four cards and if the cards are all of the same suit it is won by the side which contributes the highest card. The player (or dummy) who wins a trick is required to lead to the next trick and this rule governs much of the strategy in playing or defending a contract. When some of the cards contributed to a particular trick are in different suits the winning player is the one who played the highest card of the suit which was led, or who played the highest trump. The latter always takes precedence. A further requirement in Contract Bridge is that a player must follow suit if he can. He is allowed to Discard (play another suit) or Ruff (trump in) only when he does not have a card of the suit led.

Leading and trick-taking continue to the very end of the hand when one side will have won a number of tricks and the opponents the remainder, totalling thirteen in all. The score is then agreed and the next player in clockwise rotation deals and a new battle starts. In our example, as North dealt the first hand, East will deal the second and for that reason he will also make the first call.

No Trumps

In addition to the four suits, Spades, Hearts, Diamonds and Clubs, in descending order of rank, another designation is used in Contract Bridge called No Trumps. It is self-descriptive and means simply that there are not any trumps in this particular battle. This usually happens when one side has a preponderance of high cards but does not have a combined holding of at least eight cards in a suit, especially one of the major suits (Spades or Hearts). No Trumps enjoy the privilege of greater rank than suits, hence One No Trump as a call outranks One Spade but naturally is in turn outranked by any bid at the two level including Two Clubs, the lowest.

Valuing a hand

When a player picks up his hand and sorts it into suits he must have a method for determining whether it is wise for him to open the bidding.

There is no joy in fighting a battle on the enemy's terms and this will happen if one is unprepared. The reason is simply that the lowest bid, even if passed out, means that the partnership has to make at least seven tricks, i.e. one more than the opponents. For this reason a method of valuing hands is necessary, and the simplest one is to count 4 points for each Ace, 3 points for each King, 2 points for each Queen and one point for each Jack. Most players additionally count one point for each card in a suit over four. This is a crude method of valuation but it serves adequately as a guide. Whenever 12 points are held the hand is a border-line opening bid; it is, however, unwise to refrain from opening the bidding when holding a hand with 13 points or more.

Once a member of a partnership has opened the bidding with one of a suit, he has unlocked a door. Now his partner no longer needs 12 points to enter the fray. Knowledge that his partner has some high cards means that he should respond on as little as 6 points, i.e. half the number required for opening with a suit bid. The objective in this exchange of information is to determine which is the side's longest suit because normally that will be the one chosen as trumps.

The opponents should not be content to sit back overawed by such a show of strength. Instead they should intervene. They, too, are released from the shackles of requiring 12 points to open the bidding. Once an opponent opens, it is good tactics to interfere at the one level with a suit call whenever a good five-card or longer suit is held. The objective in this case is not constructive, but obstructive. A player has to strive to push his opponents too high for their comfort; to suggest a sacrifice to his partner; and also to assist him to find the best initial lead. These are the factors which govern overcalls – opponents must not be given a free ride or the battle will be lost.

Overcalls at the two level, however, must be up to about the strength of an opening bid with a good five-card or longer suit. An overcall of One No Trump should also be strong, about 16–18 points with a good holding in the opponent's bid suit. The reason is that to make One No Trump entails winning seven tricks. As the opponent has already indicated at least 12 points the No Trump overcaller needs to be at least the equivalent of one Ace better, hence the minimum requirement of 16 points.

Another strong interference call is known as the 'take-out double'. This call, which is simply 'Double'; indicates that the player making it has a minimum of about 12 points and ideally has support for the other suits not bid by the opponent(s). His partner is expected to respond to the double by calling his best suit, unless it has already been called by an opponent.

Scoring

Like most games, the method of scoring is important and in Contract Bridge it may, at first, seem complicated. Set out below is a simplified scoring table:–

	TRICKS MADE IN EXCESS OF SIX				
	1	2	3	4	5
No Trump =	40	70	100	130	160
Spades/Hearts =	30	60	90	120	150
Diamonds/Clubs =	20	40	60	80	100

The primary objective in rubber bridge, as its name implies, is to win the rubber. The secondary objective is to win a game because two games are required by the same side before the rubber is won. A game is when 100 or more points have been scored by declarer in making one, or more contracts. It will be seen from the above table that a contract of One Spade, if successful, i.e. at least seven tricks are won by declarer, is worth 30. To score game in Spades or Hearts on one deal therefore, ten tricks will be necessary (4 × 30). Similarly eleven tricks will be required for game if one of the minor suits, Diamonds or Clubs, is trumps. In No Trumps the first trick over six is worth 40, the second, and each subsequent trick is worth 30. This means that game can be achieved by making nine tricks in No Trumps, but only if they have been bid, i.e. by calling Three No Trumps. This condition is so important that it is worth re-stating – games, or partial scores towards game, can only be counted if a contract is bid and made.

There are penalties incurred if declarer fails to make his contract. Usually these are assessed at 50 points for each undertrick. When, however, the declarer's side already has a game the undertrick penalties are 100 each. A partnership scoring a game is said to be 'Vulnerable' because they are in greater danger of conceding a large penalty. 'Not vulnerable' therefore simply means that a partnership has, as yet, no game.

To prevent opponents from spoiling the game by outrageous overbidding, the law makers introduced the concept of 'doubling'. Effectively this increases the penalties should declarer go down in a contract. One down doubled, not vulnerable concedes 100, two down doubled is 300 and a further 200 is awarded for each additional doubled undertrick. When vulnerable, one down doubled is scored as 200 to the opponents and two down doubled 500, with 300 for each additional doubled undertrick. If a doubled contract is made, however, the declarer scores double the trick value. Four Spades, instead of being worth 120, is 240 and a further 50 is added as a bonus when any doubled (or re-doubled) contract is made.

Re-doubling, is the sanction available to declarer or his partner to deal with frivolous doubles. If successful, the trick value of the contract bid is multiplied by 4 and again the 50 bonus is added. Overtricks and undertricks when doubled/re-doubled contracts are made or defeated, are similarly increased in value.

There are additional bonuses awarded when declarer bids and makes a Small Slam (12 tricks) and a greater award for a Grand Slam (13 tricks).

In rubber bridge a further bonus is awarded to the side which wins the rubber. If by two games to nil, it is an extra 700, if by two games to one, only 500. The scoring in use in the Grand Slam match is similar to that used for rubber bridge but it has been simplified because each board is separate from each other board and therefore an adjustment in the scoring is calculated to reflect this. It is possible in rubber bridge for example to score a game with two or more part-scores, e.g. One No Trump (40) plus Two Spades (60) = 100. In Grand Slam this is not possible, hence a part-score, if successful, is awarded an additional 50 points. A non-vulnerable game is worth 300 and one that is vulnerable is worth 500.

For ease-of-reference the scoring used in the Grand Slam match is set out below in tabular form.

	NOT VULNERABLE	VULNERABLE
Bonus for part-score =	50	50
Bonus for game =	300	500

All other scores are the same as those that apply in rubber bridge with the exception that honours (four Aces in one hand in No Trump contracts, or four of the five top trump honours, or all five top trump honours, in one hand in suit contracts) do not count.

International match points

In competitions throughout the world, scoring by international match points (imps) is standard. It is simply a method whereby the net aggregate score on any one hand is converted to an imp score. The object of this exercise is to give greater weight to part-score and game contracts. In the early days of duplicate bridge when aggregate scoring was in vogue a match could be won or lost on one hand by a declarer making a lucky grand slam, despite losing small scores on most of the other hands. The imp table is set out overleaf.

15

IMP Scale	270 – 310 = 7 imps	1300 –1490 = 16 imps
	320 – 360 = 8 imps	1500 –1740 = 17 imps
0 – 10 = 0 imp	370 – 420 = 9 imps	1750 –1990 = 18 imps
20 – 40 = 1 imp	430 – 490 = 10 imps	2000 –2240 = 19 imps
50 – 80 = 2 imps	500 – 590 = 11 imps	2250 –2490 = 20 imps
90 –120 = 3 imps	600 – 740 = 12 imps	2500 –2990 = 21 imps
130 –160 = 4 imps	750 – 890 = 13 imps	3000 –3490 = 22 imps
170 –210 = 5 imps	900 –1090 = 14 imps	3500 –3990 = 23 imps
220 –260 = 6 imps	1100 –1290 = 15 imps	4000+ = 24 imps

Bidding conventions

In Contract Bridge it is necessary for partners to communicate with each other during the bidding stage. By the rules, however, they are obliged to use simply the calls: 'Pass' (or 'No Bid'), 'One' to 'Seven' of a Suit or No Trump, 'Double' or 'Re-double'. No other means of imparting information is allowed. It is also essential that a partnership understands and takes into account the bidding of their opponents. To this end bidding conventions have been invented so that a common language is available. It would be impractical in everyday life for, say, a Chinaman and an Eskimo to carry on a conversation unless they could both speak a mutual third language. This is similarly true of Contract Bridge players. In the Grand Slam contest not all the bids made will be what they seem but the commentary will explain how the players interpret these conventional calls. Similarly in the hands in this book the bidding sequences have been annotated so that readers will in effect receive a translation.

Bidding and lead conventions used by the various participants in the Grand Slam match

For convenience of readers here is a very brief outline of the bidding methods of the four partnerships competing in the Grand Slam match. The British players methods are loosely based on the Acol System and the visitors on the Standard American System.

1 No Trump (a) 12–14 points, theoretically a balanced hand with no singleton or void (Priday and Rodrigue vulnerable and not vulnerable; Gardener and Davies only when not vulnerable).

(b) 15–17 points balanced as described above (both

American partnerships and Gardener and Davies when vulnerable).

1 No Trump – 2 Clubs Response is the Stayman Convention. The opening bidder is expected to bid a major suit of four cards if holding one, otherwise to bid Two Diamonds (all players).

1 No Trump – 2 Diamonds Transfer bid, response indicates five or more Hearts. Opening bidder usually 'converts' to Two Hearts and the auction progresses, but a round of bidding has been gained. Responder, however, may be weak and pass opener's correction to Two Hearts but one advantage is that the No Trump hand remains concealed (all players).

1 No Trump – 2 Hearts Transfer bid showing spades (all players).

1 No Trump – 2 No Trump Transfer bid showing a minor suit (Granovetter and Silverman; Priday and Rodrigue).

1 Heart or 1 Spade At least 12 points and four cards in the bid suit. (The American partnerships require five cards in the bid suit). Granovetter and Silverman additionally play a forcing One No Trump response.

1 Club or 1 Diamond At least 12 points but may have only three cards in the bid suit (Gardener and Davies when vulnerable; Granovetter and Silverman; Moss and Mitchell vulnerable or not vulnerable).

1 of a suit – 2 No Trumps Strong hand, forcing (except Moss and Mitchell).

2 Clubs Conventional, very strong hand (all players).

2 Diamonds (*a*) Multi-coloured Convention – has various meanings (weak hand with a six card major suit, strong hand with a long Club or Diamond suit, 21–22 No Trump type hand, strong three suited hand (Gardener and Davies; Priday and Rodrigue).

(*b*) Weak hand with a six card Diamond suit (both American partnerships).

17

2 Hearts or 2 Spades (*a*) Strong hands usually with six cards in suit bid (both British partnerships).

(*b*) Weak hands with six cards in suit bid (both American partnerships).

(*c*) Priday and Rodrigue's 2♠ opening bids may also show a weak hand with a Club suit.

2 No Trumps Balanced hand, 19–20 (Gardener and Davies), 20–22 (other partnerships). Transfer suit responses at three level.

3 Clubs, 3 Diamonds, 3 Hearts, 3 Spades Weak hands with a seven card suit. Usually in the suit bid but Priday and Rodrigue. pre-empt in the suit below their seven-card suit.

3 No Trumps (*a*) Limited hand usually with a solid seven-card minor suit (both British partnerships).

(*b*) The American men use this bid to show a weak minor-suited hand where the suit is not solid.

4 Clubs or 4 Diamonds Goodish hand based upon a near solid seven or eight-card Heart or Spade suit, 4♣ showing Hearts and 4◇ showing Spades (all partnerships).

4 No Trumps Blackwood Convention inquiring how many Aces partner holds (Gardener and Davies). Other partnerships have slightly different methods but all use Four No Trumps as an Ace inquiry bid.

1 Diamond – 2 Diamonds or 1 Club – 2 Clubs Strong raise, forcing to at least three of the minor suit bid (all partnerships, except Priday and Rodrigue).

2 No Trumps – 3 Clubs Major suit inquiry (five cards required by Gardener and Davies before opener re-bids Three Hearts or Three Spades). All players use a form of this convention.

Defensive calls

Cue-bid of opponent's suit	e.g. One Diamond – Two Diamonds is known as the Michaels Convention indicating a weak hand with two suits usually 5–5 including unbid major(s) (all partnerships).
1 No Trump – 2 Clubs or 2 Diamonds	Overcall is conventional usually showing two or three suits (all players).
Jump overcalls in suits	(*a*) Six cards in the suit bid with a hand of about 12 points (both British partnerships).
	(*b*) Weak hand with six-card suit (both American partnerships).
Negative-type doubles	When opponents overcall, especially at the one level, i.e. the double is not for penalties, partner is expected to bid again (all partnerships).

Lead style

Top of sequence (Gardener and Davies); second highest of sequence (All other partnerships).

Ace from Ace-King (Gardener and Davies); King from Ace-King (all other partnerships).

MUD – middle-up-down and second highest from four spot cards (all partnerships).

Fourth best from long suits generally (all partnerships).

Attitude leads against No Trumps – low card denoting at least one high honour card (all partnerships).

Session One

BOARDS ONE TO TWELVE

'Boundaries before lunch? Not in the Roses Match, sir.'

The first session of a 78-Board Bridge Match is rather like the first morning of a Test Match in which it is rather more important not to lose four or five wickets than it is to score 100 runs. However, it is Nicola Gardener's and Pat Davies' style to look for scoring chances from the first deal. On Board One, they bounced out aggressively looking for a slam and settled, not altogether convincingly, for a game with a 4–3 trump fit, while the American women rejected a tentative slam invitation for the apparently obvious game. On Board Three, Gardener aggression metaphorically bruised the American men's shins and by Board Nine, the British seemed to be settling smoothly into their rhythm. Whereupon proceedings were much enlivened by Rodrigue and Priday tripping up over the bid of a three-card suit and falling on their corporate noses while Nicola Gardener inadvertently solved a riddle for Neil Silverman and helped the Americans into an excellent and unbeatable slam.

Sixty-five of the boards were dealt for the players, who were warned that a few had been slightly doctored to ensure points of analytical interest. Going by the 13 they dealt for themselves, the effort was not really necessary. However, a certain seed of doubt seemed to be sown in one or two players' minds; it is no more healthy to believe that every hand has a trap than it is to suspect your cue has turned to rubber just as you are about to pot the black.

Board One

Dealer North   N-S Vulnerable

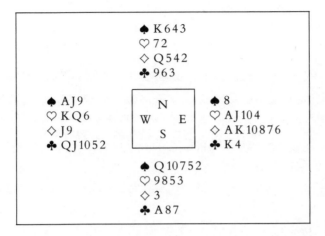

Room One

South	West	North	East
RODRIGUE, UK	MOSS, USA	PRIDAY, UK	MITCHELL, USA
–	–	NB	1◇
NB	2♣	NB	3◇
NB	3NT	All pass	

Contract: 3NT by West. Lead: ♠3.

Room Two

South	West	North	East
GRANOVETTER, USA	GARDENER, UK	SILVERMAN, USA	DAVIES, UK
–	–	NB	1◇
NB	2♣	NB	2♡[1]
NB	2♠[2]	NB	3◇[3]
NB	3♡[4]	NB	4♣[5]
NB	4◇[6]	NB	4♡[7]
All pass			

Contract: 4♡ by East. Lead: ♠5.

In Room One, Gail Moss (West) decided that in spite of her partner's forcing 3◇ rebid, the values were insufficient for a slam. She settled for what she thought would be the easiest game to make, and finished by making two over-tricks for a score of 460 to the USA. In Room Two, the English women agonised a little.

1 DAVIES: 'I'm slightly low on points to reverse into 2♡ here, but on the other hand, I'm much too good in terms of distribution just to rebid 2◇. And I quite like my doubleton ♣K when Nicola has bid 2♣.'

2 GARDENER: 'So partner has about a 16 count and is 5 – 4 in Diamonds and Hearts. We must be close to a slam because the high cards I have outside Clubs are very good. I'm not going to pre-empt things and decide to play 3NT. I'll bid the fourth suit and see how Pat reacts.'

3 DAVIES: 'Well, she wants to know more about my hand. I've shown her five Diamonds and four Hearts; now's the chance to show her the sixth Diamond.'

4 GARDENER: 'It looks as if she's 6 – 4 – 2 – 1, and if she has one Spade and two Clubs, her hand is very unattractive in 3NT. She hasn't got solid Diamonds because if she had had AKQ to six, she would have started the bidding with one of our Two Diamond bids. There must also be a gap in the Club suit so I'm rather discouraged from bidding 3NT now. I'll make a waiting bid and see what she does.'

5 DAVIES: 'Murky waters here. I'm not sure whether we're looking for the best game or the best slam. I'd like to show her my ♣K; it means by-passing 3NT, but I don't think 3NT can be on if Nicola can't bid it.'

6 GARDENER: 'Now, I can show the Diamond support and see what she does.'

7 DAVIES: 'Still not really sure where we're going. If I bid 4♡, she'll know I've only got four of them, and if she wants to pass, she can.'

Nicola Gardener did pass, and Pat Davies was left in a situation which causes less experienced players to perspire so freely – how to make a contract with only seven trumps in the two hands. Pat Davies: 'I can either try to establish a side suit, draw trumps and make the tricks in the side suit, or I can try and make ten tricks on a cross-ruff. The problem with establishing a side suit is that I may have trouble getting back into that hand to cash the tricks. In a cross-ruff, I suppose I could be over-ruffed, but on the other hand, my trumps are so good I can ruff high and come to ten tricks that way. I'll win with the ♠A straight away because hopefully I'm going to be ruffing Spades in one hand and Diamonds in the other.'

Which is what happened, Pat Davies making to ten tricks and a score of 420 to the UK, giving a gain of 1 imp to the USA.

Interestingly, even though there are only 29 high-card points in the East-West hands, either Six Clubs or Six Diamonds is on. In Six Diamonds, it looks as if the right play would be to cash the ◇A in East's hand to guard against the singleton ◇Q being in the South hand. But this would be poor play, quite against the odds. Correct is to finesse the Diamonds on the first round, otherwise with only two Diamonds in the West hand the Queen to four in the North hand can't be subjected to a second successful finesse.

Board Two

Dealer East   N-S Vulnerable

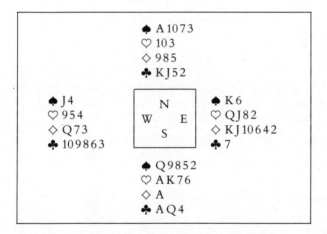

Room One

South	West	North	East
RODRIGUE, UK	MOSS, USA	PRIDAY, UK	MITCHELL, USA
–	–	–	NB
1♠	NB	2♠	3◇
4◇	NB	5♣	NB
6♠	All pass		

Contract: 6♠ by South. Lead: ◇3.

Room Two

South	West	North	East
GRANOVETTER, USA	GARDENER, UK	SILVERMAN, USA	DAVIES, UK
–	–	–	1◇
Dble	1NT	2♠	NB
3◇	NB	4♠	NB
4NT	NB	5◇	NB
6♠	All pass		

Contract: 6♠ by North. Lead: ♣7.

The first slam of the match, precisely bid and played in both rooms for no swing. The interesting point of the hand is the play of the Spade suit. There is nothing wrong in playing the ♠A from the North hand and then low towards the Queen. A good psychological way of playing the hand, though, is to lead the ♠3 from North towards the Queen, putting the pressure on East to duck with King and another.

Board Three

Dealer South   E-W Vulnerable

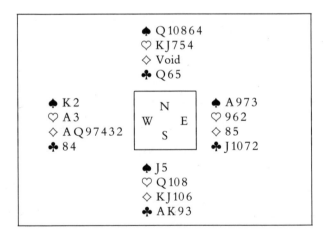

♠ Q10864
♡ KJ754
◇ Void
♣ Q65

♠ K2
♡ A3
◇ AQ97432
♣ 84

♠ A973
♡ 962
◇ 85
♣ J1072

♠ J5
♡ Q108
◇ KJ106
♣ AK93

Room One

South	West	North	East
RODRIGUE, UK	MOSS, USA	PRIDAY, UK	MITCHELL, USA
1♣[1]	1◇[2]	1♠[3]	NB
1NT[4]	2◇[5]	3♡[6]	NB
3NT[7]	All pass		

Contract: 3NT by South. Lead: ◇2.

1 RODRIGUE: 'Tony and I play a 12–14 No Trump. Superficially I've got 14 points but it's a jolly good 14 with two Tens and a Nine. I think it's too good for a No Trump so I'll promote it to 15 points and open a Club.'

2 MOSS: 'I'd really like to bid a large number of Diamonds because I don't have much defence against a game in either major. But if I did that, it would show a weak hand, and my hand's too good to pre-empt. I shall have to forego discomfitting the opponents in order to be accurate.'

3 PRIDAY: 'Not many points, but a nice shape.'

4 RODRIGUE: 'I can now show my 15–17 points (!)'

5 MOSS: 'Why is it my lovely hands are always diminishing before my eyes? I wanted to bid a lot of Diamonds before, but my hand was too good. Now I still want to bid them, but it is not good enough. I can't be a coward, though; and I do have seven of them.'

6 PRIDAY: 'If I simply show my second major suit at this point by bidding 2♡, Claude could pass, and I certainly can't risk that bearing in mind I've got quite a good fit in his first bid suit. I'm just going to make sure we reach game.'

7 RODRIGUE: 'So, partner's got five Spades and four Hearts. I've got a very good double guard in Diamonds, so from this side of the table it looks as if the obvious contract should be 3NT.'

It looked an even better contract from the Americans' point of view. After winning the first Diamond trick, Rodrigue got off lead with the ♡Q, won by West who cashed the ♠K, led a small Spade to her partner's Ace and made two Diamond tricks on the return of the ◇5 to set the contract one trick. Plus 50 to the Americans. Meanwhile . . .

Room Two

South	West	North	East
GRANOVETTER, USA	GARDENER, UK	SILVERMAN, USA	DAVIES, UK
1♣	2◇	Dble[1]	All pass

Contract: 2◇ Doubled by West. Lead: ♣5.

In Room Two, Matthew Granovetter sitting South opened a routine 1♣ for the Americans. Nicola Gardener suffered from none of Gail Moss' inhibitions and bid 2◇ hoping her partner might have some values in Clubs and that they might go looking for a game. And that bid fixed Neil Silverman properly.

1 SILVERMAN: 'I have two bids here and unfortunately, they are both bad. I could bid Two Spades and then bid my Hearts later, but that might all get too high. My other choice is to make a Negative Double, showing the major suits. But usually, one only has four of one, let alone five of both. I suppose I could pass, but my hand has very little defensive value. I don't think I should force to game with this hand. I'll just take a chance my partner will bid.'

But sitting with four trumps to the King and Jack, Granovetter was very happy to let Nicola Gardener struggle, and probably go down, in 2◇ Doubled. Unfortunately for the Americans, she made it (+ 180) for a net plus score of 130 to the United Kingdom and a gain of 4 imps on the board, which re-inforces the old maxim: 'Be wary of doubling for penalties when your trumps are *under* declarer'.

Board Four

Dealer West   Game All

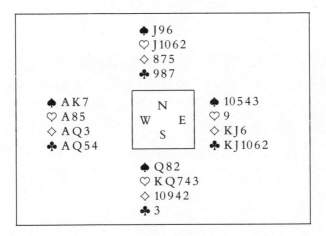

♠ J96
♡ J1062
◇ 875
♣ 987

♠ AK7
♡ A85
◇ AQ3
♣ AQ54

N
W E
S

♠ 10543
♡ 9
◇ KJ6
♣ KJ1062

♠ Q82
♡ KQ743
◇ 10942
♣ 3

Room One

South	West	North	East
RODRIGUE, UK	MOSS, USA	PRIDAY, UK	MITCHELL, USA
–	2♣	NB	2◇
NB	2NT	NB	3♣
NB	3◇	NB	4♣
NB	4◇	NB	6♣
All pass			

Contract: 6♣ by West. Lead: ♣8.

Room Two

South	West	North	East
GRANOVETTER, USA	GARDENER, UK	SILVERMAN, USA	DAVIES, UK
–	2♣	NB	3♣
NB	4♣	NB	4◇
NB	4♡	NB	5♣
NB	5♠	NB	6♣
All pass			

Contract: 6♣ by West. Lead: ♠6.

There is nothing to the play of this hand; do what he will, declarer must always lose just one Spade trick. The players themselves thought the bidding was so routine that they could see no point in giving their thoughts, but as an exercise, even if you have read the bidding, cover the N-S cards, go to your favourite partner and see what contract you end in with the E-W cards. If you reach 6♣, Neil Silverman will not want to play blindfold against you.

Board Five

Dealer North   Love All

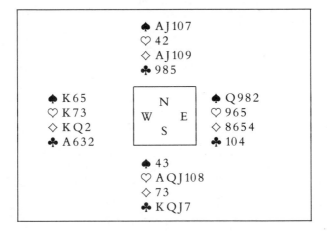

Room One

South	West	North	East
RODRIGUE, UK	MOSS, USA	PRIDAY, UK	MITCHELL, USA
–	–	NB	NB
1♡	Dble	Redble	1♠
NB	NB	Dble	All pass

Contract: 1♠ Doubled by East. Lead: ♣Q.

Room Two

South	West	North	East
GRANOVETTER, USA	GARDENER, UK	SILVERMAN, USA	DAVIES, UK
–	–	NB	NB
1♡	Dble	Redble	1♠
NB	NB	Dble	All pass

Contract: 1♠ Doubled by East. Lead: ♠3.

Nothing dramatic about it, but a curious little hand nonetheless made the odder by the fact that the same inferior contract was reached in both rooms. Over East's rescue attempt of 1♠, South, who had originally opened the bidding, left the final decision to his partner. Tony Priday, sitting North in Room One, had no illusions about the difficulties he faced: 'I know Claude has an opening bid and that means Jackie can only have very few points. So I have the choice of three calls. I can pass and hope to take a good penalty or I can bid one or two No Trumps showing that I have the Spades well held and good values. It's certainly too strong for One No Trump. Perhaps Two No Trumps would be about right, but we're not vulnerable and at Love All I think I shall try and take a good substantial penalty and teach these girls what we are made of.'

In Room Two, the American North, Neil Silverman, was more succinct about Pat Davies' chance of making One Spade: 'I don't think they're gonna make this, especially if my partner has some values, which he sort of promised. I'll just see how much we can beat them.'

In fact, on the lead of a trump, they defeated the contract by two tricks for a penalty of 300 and then pointed out they should have beaten it by another trick for a penalty of 500 which would have been an excellent result, assuming the British men in Room One had reached and made the correct contract of 3NT on the North-South cards. However the British men did not bid Three No Trumps, nor excel in defence, aiming for a ruff and finishing by beating the contract by only one trick for a net gain of 200 points and 5 imps to the Americans. Certainly Granovetter's trump lead seems a better choice than Rodrigue's ♣Q.

And so to Board Six where both East-Wests played quietly in 3◊, going one down, to leave the Americans with a two imp lead – six to four – after the break.

Board Six

Dealer East   N-S Vulnerable

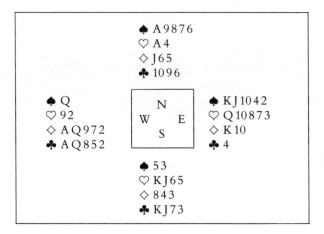

```
                    ♠ A9876
                    ♡ A4
                    ◇ J65
                    ♣ 1096
   ♠ Q            ┌─────────┐      ♠ KJ1042
   ♡ 92          │    N    │       ♡ Q10873
   ◇ AQ972       │ W     E │       ◇ K10
   ♣ AQ852       │    S    │       ♣ 4
                  └─────────┘
                    ♠ 53
                    ♡ KJ65
                    ◇ 843
                    ♣ KJ73
```

Room One

South	West	North	East
RODRIGUE, UK	MOSS, USA	PRIDAY, UK	MITCHELL, USA
–	–	–	NB
NB	1◇	1♠	NB
NB	2♣	NB	2NT
NB	3♣	NB	3◇
All pass			

Contract: 3◇ by West. Lead: ♡A.

Room Two

South	West	North	East
GRANOVETTER, USA	GARDENER, UK	SILVERMAN, USA	DAVIES, UK
–	–	–	NB
NB	1◇	NB	1♠
NB	2♣	NB	2♡
NB	3♣	NB	3◇
All pass			

Contract: 3◇ by West. Lead: ♡4.

Board Seven

Dealer East   E-W Vulnerable

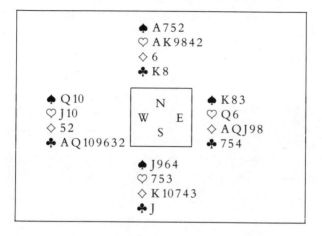

```
                    ♠ A752
                    ♡ AK9842
                    ◇ 6
                    ♣ K8
        ♠ Q10          ┌─────────┐        ♠ K83
        ♡ J10          │    N    │        ♡ Q6
        ◇ 52           │ W     E │        ◇ AQJ98
        ♣ AQ109632     │    S    │        ♣ 754
                       └─────────┘
                    ♠ J964
                    ♡ 753
                    ◇ K10743
                    ♣ J
```

Room One

South	West	North	East
RODRIGUE, UK	MOSS, USA	PRIDAY, UK	MITCHELL, USA
–	–	–	1◇
NB	2♣	Dble	NB
2♠	NB	3♠	All pass

Contract: 3♠ by South. Lead: ◇5.

Room Two

South	West	North	East
GRANOVETTER, USA	GARDENER, UK	SILVERMAN, USA	DAVIES, UK
–	–	–	1◇
NB	2♣	Dble[1]	NB[2]
2♠[3]	3♣[4]	3♡[5]	4♣[6]
NB[7]	NB	NB[8]	

Contract: 4♣ by West. Lead: ♡K.

The first board of the second half of the session showed the British team at its best and worst. Away from the table, Nicola Gardener and

31

Pat Davies do not have the natural arrogance of World Champions; they both radiate a touch of vulnerable diffidence, but at the card table it is a different matter. Any Barbara Cartland hero who tried, in the course of battle, to fold either of them in his warm, protective embrace would find himself short-arm jabbed, probably rabbit punched and certainly thrown flat on his back. Nicola Gardener and Pat Davies, as Captain Najocks said of his hired sportsmen, play hard games jolly hard.

Tony Priday, on the other hand, assured, amusing, apparently in control of his own and his partner's destiny, is given to moments of diffidence that both his opponents and friends find utterly charming but which his enemies and sometimes even his partners castigate as rank timidity. Not that Claude Rodrigue would criticise his partner in public, particularly early in the match; he is far too aware of how tender a plant partnership understanding can be, even though at rubber bridge he is not always known for silent forgiveness of his partner's imbecilities.

Board Seven was not the greatest triumph for the men of either team. The Americans, perhaps still jet-lagged or not really believing their opponents would be so punchy early in the match, allowed the British women to sacrifice in 4♣. It went one off, but that would have been an excellent result if the British men in the other room had reached the aggresive contract of 4♡. As it was, they pottered diffidently into a part score and gained just one instead of possibly eight imps for the UK.

So how was it neither North-South found themselves basking in 4♡?

Room Two

1 SILVERMAN: 'My opponents seem to have most of the high cards, but if my partner has the right cards, I could almost definitely make 4♠ on this hand. I might make a take-out Double here, then bid Hearts to show partner I do have Spades and Hearts and a good hand.'

2 DAVIES: 'I'd like to make it difficult for Matthew to get into the bidding by bidding 3♣ so that he has to come in at the three level. Against this, my hand really is a minimum and Nicola might have bid 2♣ on a four-card suit.'

3 GRANOVETTER: 'I didn't think I'd ever get into this bidding; now I have to bid.'

4 GARDENER: 'Well, my hand hasn't improved at all, but I do have a seven-card suit and I wouldn't mind defending at the three level because I have a doubleton in my partner's suit and I might get a ruff.'

5 SILVERMAN: 'So far things are going according to plan.'

6 DAVIES: 'The same problem again; shall I raise partner's Clubs or shall I defend their contract? I think I'll push them and hope if they go one higher we can beat their contract.'

7 GRANOVETTER: 'Gee, this is very close. I have three Hearts so we might make 4♡ if everything is going nicely. With Diamonds bid on my right, the ◇K looks like a good card. But I had better just use a bit of caution and leave it to my partner.'

8 SILVERMAN: 'Doesn't look as if Matthew has what I hoped for, as if he had, I'm sure he would have tried 4♣ or maybe even 4♡ with some Hearts. It doesn't look as if 4♣ is going to make. I'll probably make two Hearts, a Spade and a Club. There are good chances; I'll just let them play it.'

Well reasoned comment by Silverman and, in fact, that is what he did make to set the contract one trick, but subsequent analysis showed it could have been a poor result for the Americans because 4♡ can be made by good play unless East-West are perceptive enough to attack Spades in time. Declarer eliminates Clubs from the North and South hands and extracts West's Diamonds. When he gets off play with ♠A and another to the Jack, the defence is forced to concede a ruff and discard, or sacrifice a Spade trick. If East foresees this line of play, then after winning the ◇A he leads a Spade immediately.

In Room One, Hearts were not mentioned at all. After his partner's bid of 2♠, Tony Priday thought: 'I've already shown that I've got Hearts (but not that he had six of them) and I have a choice at this moment of playing quietly in 2♠ unless opponents compete, or pushing slightly for game. Well, I like my shape and the fact that the ♣K seems to be sitting well, so I shall just give Claude one more chance by raising him in Spades.'

Rodrigue decided 3♠ was enough, and nine tricks were made for a net swing to the UK of 40, or 1 imp.

Board Eight was one of those indeterminate little part-score hands that come up so often in Bridge. In Room Two, the Americans finished in 3♣ – North's bid of 2NT over his partner's opening bid was a Transfer Bid demanding 3♣ from South. Predictably, this went one down. In Room One, the British finished in 2♠, which should have gone one down as well. However, Gail Moss sitting West chose to lead the ◇4 against South's contract and after that, Rodrigue had little difficulty in making eight tricks. Sitting with ♠KQJ5 behind Declarer, the lead of the ♠K not only does not look obvious; it appears to throw away a certain trump trick. Which indeed it does. But the trick comes back with interest, because if West does lead the ♠K, the time will come

33

when South cannot ruff his red suit losers in dummy and must go one down.

The result of this defensive slip, 3 imps to the UK, meant they regained the lead for the second time in the match.

Board Eight

Dealer South   Game All

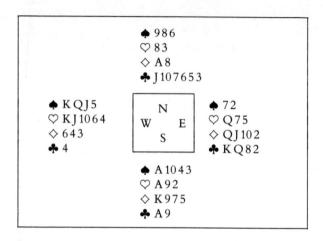

Room One

South	West	North	East
RODRIGUE, UK	MOSS, USA	PRIDAY, UK	MITCHELL, USA
1♠	NB	2♠	All pass

Contract: 2♠ by South. Lead: ◇4.

Room Two

South	West	North	East
GRANOVETTER, USA	GARDENER, UK	SILVERMAN, US	DAVIES, UK
1NT	NB	2NT	NB
3♣	All pass		

Contract: 3♣ by South. Lead: ♠K.

Board Nine

Dealer West   Love All

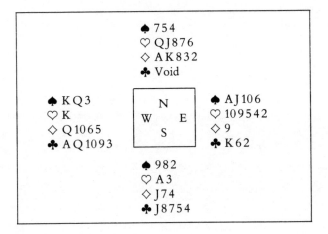

Room One

South	West	North	East
RODRIGUE, UK	MOSS, USA	PRIDAY, UK	MITCHELL, USA
–	1◇	1♡	Dble
NB	2♣	All pass	

Contract: 2♣ by West. Lead: ♠5.

Room Two

South	West	North	East
GRANOVETTER, USA	GARDENER, UK	SILVERMAN, USA	DAVIES, UK
–	1♣	1♡	Dble
NB	2◇	NB	3♣
All pass			

Contract: 3♣ by West. Lead: ♡6.

In Room One, the American women made two over-tricks; in Room Two, the British made their contract exactly for a swing of 1 imp to the USA. And so to Board Ten and the first real fireworks in the match.

Board Ten

Dealer North   E-W Vulnerable

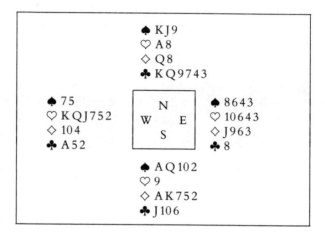

Room One

South	West	North	East
RODRIGUE, UK	MOSS, USA	PRIDAY, UK	MITCHELL, USA
–	–	1♣	NB
1♦[1]	1♡[2]	1♠[3]	NB
2♡[4]	Dble	NB[5]	NB
3♠[6]	NB	4♠[7]	All pass

Contract: 4♠ by North. Lead: ♣8.

1 RODRIGUE: 'We're going to game of course, but there's no need to force. I can keep the bidding open for a long time, and I want to describe this hand, so I'll start off by bidding my five-card suit.'

2 Moss: 'Shall I bid 1♡ or 2♡? I'd rather bid 2♡, but at this vulnerability (E-W are vulnerable, N-S are not) it could be suicidal. I'll just get across the message I'd like Jackie to lead a Heart if she's on lead.'

3 PRIDAY: 'That's a very annoying bid. Normally, I would have rebid 1NT to show a balanced hand with 15–17 points. That doesn't look very attractive now because in No Trumps I shall certainly get a Heart lead and I only have one stopper. 2♣ seems a little cautious and 3♣ is clearly excessive. There is an alternative here which I find quite attractive – and that's to bid my three card suit.'

4 RODRIGUE: 'What sweet music. Partner is obviously not minimum, and I've got a marvellous fit in his second suit. I've got second round control of the opponent's suit. I've got a good five-card suit and I've also got useful fitting cards in Tony's opening suit. Let's get in the good news straight away and tell him I've got a Heart control.'

5 PRIDAY: 'Dear oh dear. I'm not sure I like the way this hand is going. I might well regret that 1♠ bid at the end of the day. I suppose I could redouble to show first round Heart control, but I think I'll just wait and see what partner has in mind.'

6 RODRIGUE: "I could bid 2♠ which would be forcing and show four-card support. But in case Tony is worried and has a good hand with only King to four little Spades, I want to make it quite clear to him that I've got really very good trump support apart from a good hand. I think I'll make that clear.'

7 PRIDAY: "I feared the worst. It's clear that Claude has a big fit with me in Spades and my 1♠ bid has brought all this down on my head. Strictly speaking, I should cue bid my ♡A but I'm still worried about my Spade holding, having only a three-card suit. I think I had better be cautious and settle quietly for game.'

Even with an over-trick, that was a horrible result for the British men, rubbed in by what happened in Room Two where the Americans did reach the slam; although not without a little help from their opponents.

Room Two

South	West	North	East
GRANOVETTER, USA	GARDENER, UK	SILVERMAN, USA	DAVIES, UK
–	–	1NT	NB
2♣[1]	2♡[2]	3♣[3]	3♡[4]
4♡[5]	NB	4♠[6]	NB
4NT[7]	NB	5♠	NB
6♣[8]	All pass		

Contract: 6♣ by South. Lead: ♡K.

1   Stayman, trying to find out if they have a fit in the majors.

2 GARDENER: 'To bid or not to bid? We're vulnerable and it could be dangerous. On the other hand, I do want a Heart lead if they get to 3NT. I don't know. My father always used to say, if you have a six-card major suit, bid it.'

3 SILVERMAN: 'Sometimes you get lucky. If Nicola hadn't bid, 3♣ by me would have shown both majors. Now it will just show six Clubs.'

4 DAVIES: 'It's not often I support partner on a one count, but Nicola has bid at the two level, vulnerable against not vulnerable, so I'll show the support I have for her.'

5 GRANOVETTER: 'What is this? What are they doing in our auction – have they no respect for us? It sure looks like my partner has good values outside of Hearts the way they keep bidding them on my left and right. I'd say there was 6♣ on here, but how do I raise my partner? If I bid 4NT right now, maybe it'll sound quantitative and if I bid 4♣, maybe that'll sound like I'm just competing. I need a cue bid that will show him Club support. 4♡; and from there it will go at least to 6♣.'

6 SILVERMAN: 'So partner likes his hand and is interested in a slam. I'd better show where my values are.'

7 GRANOVETTER: 'I'll try key-card Blackwood.'

8 GRANOVETTER: 'That shows two key cards, the ♡A and the ♣K plus the Queen of trumps since 5♡ would have shown two key cards without the Queen of trumps. Oh we're missing an Ace then. I'll sign off in a small slam.'

Not the easiest of slams to bid, but an excellent contract with only the ♣A to lose.

If Board Ten was a nightmare for the British, Board Eleven might have been specially devised to torment the women. A truly horrible trump break, and not so easy to see how to make ten tricks even when all 52 cards are exposed.

Board Eleven

Dealer East   Game All

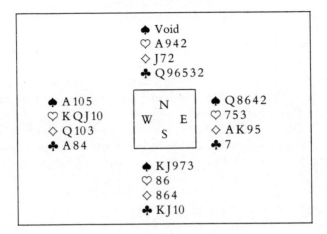

Room One

South	West	North	East
RODRIGUE, UK	MOSS, USA	PRIDAY, UK	MITCHELL, USA
–	–	–	NB
NB	1NT	NB	2♡
NB	2♠	NB	3NT
NB	4♠	All pass	

Contract: 4♠ by West. Lead: ♣5.

Room Two

South	West	North	East
GRANOVETTER, USA	GARDENER, UK	SILVERMAN, USA	DAVIES, UK
–	–	–	NB
NB	1NT	NB	2♡
NB	2♠	NB	3♢
NB	4♠	All pass	

Contract: 4♠ by West. Lead: ♣3.

After both Easts had used a Transfer Bid to show five Spades, neither partnership had much difficulty in finishing in 4♠. More interesting was

39

that both Souths had the self-control not to double and give away the awful trump break. Silverman, the American North, recalled that in his youth he used to double this sort of contract to protect his partner's four or five-card Spade holding, but then as he got older and his opponents started re-doubling and making an overtrick, he gave it up and learned to pass.

Nicola Gardener decided she could not make the contract if the Spades broke five nothing, so after winning the first trick, she led a small one towards dummy's Queen. Gail Moss toyed with the idea of crossing to dummy to lead up to the ♠A 10, but rejected it for fear of a Diamond ruff, and so both East-Wests went one down for no swing on the board.

As the cards lie if Gail had risked crossing to dummy with a Diamond in order to lead a low Spade to the Ten, she could have made this contract. When the 5–0 trump break is discovered she is forced to ruff Clubs in dummy, taking care to set up one Heart trick in the process. In the end position after declarer has made eight tricks, South has to ruff and return a trump allowing declarer to make dummy's Queen for her ninth trick. The ♠A is the tenth. It does not help if South inserts the ♠J on the first round, the end position is similar.

Board Twelve was a simple hand on which both Norths opened 1♠, both Souths responded 1NT where they played, going one down.

So, at the end of the first session, thanks to the big swing on Board Ten, the Americans led by 17 imps to 10.

Board Twelve

Dealer South   E-W Vulnerable

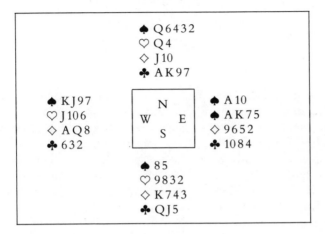

Room One

South	West	North	East
RODRIGUE, UK	MOSS, USA	PRIDAY, UK	MITCHELL, USA
NB	NB	1♠	NB
1NT	All pass		

Contract: 1NT by South. Lead: ♡J.

Room Two

South	West	North	East
GRANOVETTER, USA	GARDENER, UK	SILVERMAN, USA	DAVIES, UK
NB	NB	1♠	NB
1NT	All pass		

Contract: 1NT by South. Lead: ♡J.

Session Two

BOARDS THIRTEEN TO TWENTY-FOUR

'You mean to say she doubled? Pat Davies doubled? Good grief.'

With the men playing the men and the women playing against the women, a dreadful session for the Americans saw them lose 68 imps and win just 12, and yet it all began so well. On the first Board, Pat Davies stretched her values a little to bid an unmakeable slam, only for Gail Moss inadvertently to punish her partner on the next Board and give nine imps back to the British. That was only a prelude to Board Fifteen, where the British women bid and made a slam while in the other room, Rodrigue and Priday were robbing Silverman and Granovetter of their inheritance.

The session continued with a series of misjudgements from the Americans. Granovetter sprung anti-system bids on two successive Boards, Gail Moss got busy when she should have played passive, and then came one of the most unexpected bids in the game. Sitting with three trump tricks and an outside Ace, Pat Davies passed 3♡ and then to her astonishment heard the Americans bid 4♡. Heartened by their successful double, the British women then found an inspired defence on the next hand, while Claude Rodrigue was demonstrating the technique that makes him one of the finest dummy players to rake in another 12 imps for the British. On Board Twenty-two a defensive misjudgement by Jackie Mitchell combined with imperfect play by Granovetter to give the British a further ten imps and the session ended with the British having turned a deficit of seven into a lead of 49.

Board Thirteen

Dealer South   Love All

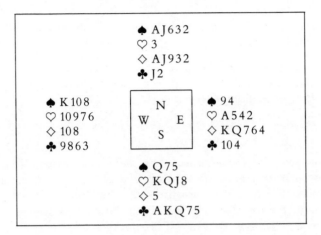

Room One

South	West	North	East
DAVIES, UK	MOSS, USA	GARDENER, UK	MITCHELL, USA
1♣¹	NB	1♠	NB
2♡	NB	3◇²	NB
3♠³	NB	4◇⁴	NB
4♡⁵	NB	5♠⁶	NB
6♠	All pass		

Contract: 6♠ by North. Lead: ♡A.

Room Two

South	West	North	East
GRANOVETTER, USA	RODRIGUE, UK	SILVERMAN, USA	PRIDAY, UK
1♣	NB	1♠	NB
2♡	NB	3◇	Dble
3♠	NB	4♣	NB
4♠	All pass		

Contract: 4♠ by North. Lead: ◇Q.

1 DAVIES: 'One of my better looking hands. I can open with Clubs and reverse in Hearts showing both suits and five Clubs.'

43

2 GARDENER: 'She must have five Clubs and four Hearts. I wonder whether we have a fit in Spades or Diamonds or if No Trumps should be the spot. I can introduce the fourth suit and find out more about her hand.'

3 DAVIES: 'Now I can show Spade support and partner will realise I am short in Diamonds.'

4 GARDENER: 'My hand's looking better and better. We've got a Spade fit; my hand is rich in controls and we could be heading for a slam. I'll cue bid the ◇A and see what Pat does.'

5 DAVIES: 'So she agrees Spades and has the ◇A. This hand looks quite attractive now. I can show a Heart control; either A or K.'

6 GARDENER: 'That's interesting. She could have signed off in 4♠ but it sounds as if she's got a bit to spare, and as I've got a bit to spare as well I'd hate to sign off in 4♠ when we could have Six on. Really, I'd like to invite the slam but not bid it. I can show I am not worried about the Heart or Diamond control; I'll just make a value bid and see what she does.'

7 DAVIES: 'So, I'm being invited to bid Six if I've got good enough Spades. The question is what constitutes good Spades in this auction? Queen to three wouldn't be good if I had bid the suit myself or if I'd raised her immediately, but as I didn't support them until my third bid I think Queen to three might be as much as she is expecting. It's close, but I think my Spades will be what she wants.'

Not the most desirable of contracts. North had to lose the ♡A and the only distribution that would have allowed her not to lose a Spade would have been exactly King and one other in the West hand. The bid that caused the problem was Pat Davies' 4♡. Not having the Ace and with nothing in reserve after her reverse she should have signed off in 4♠.

The American men bid to a pedestrian 4♠, made the same eleven tricks, gained 11 imps on the board and stretched their lead to 18 imps.

Board Fourteen

Dealer West   N-S Vulnerable

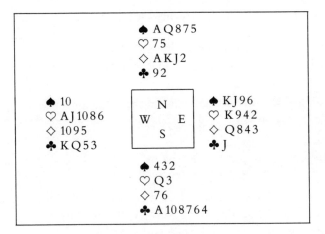

Room One

South	West	North	East
DAVIES, UK	MOSS, USA	GARDENER, UK	MITCHELL, USA
–	NB	1♠	NB
2♠[1]	Dble[2]	3◇[3]	3♡[6]
NB[5]	4♡[6]	All pass	

Contract: 4♡ by East. Lead: ◇7.

Room Two

South	West	North	East
GRANOVETTER, USA	RODRIGUE, UK	SILVERMAN, USA	PRIDAY, UK
–	NB	1♠	NB
2♠	Dble[1]	Redble[2]	3♡[3]
NB[4]	NB	3♠[5]	NB[6]
NB	NB		

Contract: 3♠ by North. Lead: ♣J.

This was an intriguing little board in which the American women found themselves playing in 4♡ for no very obvious reason and going two down on the E-W cards, while in the other room the American men sitting N-S bought the contract in 3♠ which was utterly doomed,

45

and should have cost them 13 imps. That it only cost them nine imps could be attributed to what Tony Priday's friends would call his natural diffidence and his critics on a polite day would term his caution.

Room One

1 DAVIES: 'I've got a six-card Club suit, but in our system a bid at the two level shows a good eight or nine points, so I'm not good enough to bid 2♣.'

2 MOSS: 'Well, I'm a passed hand, we're not vulnerable and they are, so now is the time to show partner I can support any of the other three suits.'

3 GARDENER: 'I've got a good 3♠ bid, but why not bid 3◇ and give partner an idea of where our defensive tricks are?'

4 MITCHELL: 'I don't see why I can't compete with this hand. I've got four Hearts and I'm sitting over North with four Spades.'

5 DAVIES: 'I'm rapidly going off this hand. I've got the worst three Spades in the pack; no points in either of my partner's suits so I'd rather defend.'

6 MOSS: 'My hand certainly doesn't justify any more bidding. I know we'd be lucky to make 3♡, but Nicola's made a game try so it's certainly going to be close. With my five-card Heart suit and partner bidding Hearts, we certainly don't seem to have much defence.'

And no chance of making that contract against accurate defence. North cashed her two top Diamonds and the Ace of Spades, gave her partner a Diamond ruff and had the joy of seeing partner cash the ♣A to set the contract two tricks. Moss must surely take the blame. Her 4♡ bid punishes her partner for showing some initiative.

Room Two

1 RODRIGUE: 'I could come in with Hearts here, but Double is a far better call because partner could be short of Hearts and have a fit in one of the minor suits. He could even conceivably pass a Double of 2♠ for penalties. He can't expect any more from me than I have.'

2 SILVERMAN: 'These Britishers; so polite and always in our auctions. This time I think he might have made a mistake. It's possible that besides Spades, partner has Hearts and Clubs and it'll be easier for him to double them if I show some values. Apart from that, if I redouble, he knows that we might have a possible game.'

3 PRIDAY: 'Well, he's passed, but Claude must have a shapely hand with a few points. I have a Heart suit; he'll certainly have Hearts, so I'm well placed despite the redouble.'

4 GRANOVETTER: 'My hand has gotten worse with Queen and one Heart. I'm passing here and hoping Neil passes also.'

5 SILVERMAN: 'This didn't work out too well. The last time I bid 3♠ in this sort of position, I got socked for 500 but I'm going to do it again.'

6 PRIDAY: 'I'd like to double this, but I'd be punishing Claude if I did. He tends to re-open on rather light, distributional hands, and anyway if I were to double it would give away where the Spades are. Not that they don't know that already, but I might find some good honour Spade on my left, so I'm certainly going to pass and hope we do well on the hand.'

Perhaps if Silverman encounters the situation for the third time he will pass as he undoubtedly should have done here. Ah well, there are more ways of winning a match than taking the points that seem to be on offer. Plus nine to the United Kingdom. And now to something really dramatic. Before looking at how the experts handled Board Fifteen, cover up the bidding and see what you would expect the final contract to be.

Board Fifteen

Dealer North   Game All

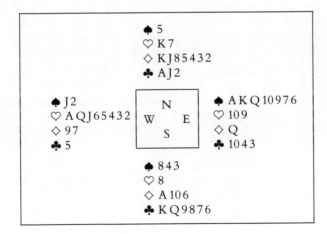

Room One

South	West	North	East
DAVIES, UK	MOSS, USA	GARDENER, UK	MITCHELL, USA
–	–	1◇	4♠
4NT¹	5♡²	6◇³	All pass

Contract: 6◇ by North. Lead: ♠K.

1 DAVIES: 'That makes life difficult. If she had bid Spades at a lower level, I'd certainly have shown my six-card Club suit. The question is; can I go on to the five level opposite an opening bid from partner? We play 4NT in this sort of situation to show some sort of tolerance for her Diamond suit and an outside Club suit. In other words, I can ask her to pick which minor she thinks we should play in. It's not often we get the chance to use that bid, but now's a good time.'

2 MOSS: 'What a predicament. I certainly have good Spade support but if I bid 5♠ now, how will my partner know that I would like her to lead a Heart? I think perhaps I should mention the Hearts now and conceivably I'll be in a better position to decide who should be sacrificing against whom by the next go.'

3 GARDENER: 'I hate these high level decisions. I never know what to do. I don't know whether I am bidding it to make, or sacrificing, but it certainly looks right to plough on in Diamonds.'

Of course 6◇ should go one down. East cashes the ♠K and will then lead a Heart, the suit her partner has bid, for the defence to cash their second trick. But what is this? An agonising trance from Jackie Mitchell at trick two. She lays her head down on the table. She sits up. She half draws a card from her hand, then pushes it back. Draws it again, puts it back and with the nearest a professional would allow herself to a deep sigh, leads the ♠A! Declarer ruffs, draws trumps and discards her losing Hearts. To attempt to explain Jackie Mitchell's trance would be to invite retribution; so we will leave them to analyse behind closed doors. . . . And while the British women were bidding and making an impossible slam, the British men were adding to the Americans' discomforture.

Room Two

South	West	North	East
GRANOVETTER, USA	RODRIGUE, UK	SILVERMAN, USA	PRIDAY, UK
–	–	1♦	2♠[1]
3♣[2]	4♡[3]	NB[4]	4♠[5]
NB[6]	NB[7]	4NT[8]	NB[9]
5♣[10]	NB[11]	NB[12]	5♡[13]
NB	NB	NB[14]	

Contract: 5♡ by West. Lead: ♣A.

1 PRIDAY: 'I don't really fancy jumping to game at this vulnerabilty, but it does seem a little pusillanimous to bid just 1♠. 2♠ shows a slightly stronger hand than I have, but my Spades really can't be any better, so I'll try it.'

2 GRANOVETTER: 'He's giving me a head-ache. I have to bid my Clubs; I just have to even though it might propel us way too high. I have KQ six times and we can easily have a game; on the other hand, since Tony has an intermediate two bid, we could easily have nothing. But I must risk it.'

3 RODRIGUE: 'There seems to be a lot of distribution around on this hand. Tony has shown six Spades and values for an opening bid. What I want to do is to get the message across quite clearly that I've got a self-supporting Heart suit in spite of my partner's Spades; and also I can take up a bit of bidding space.'

4 SILVERMAN: 'I don't have an idea what to do here. Luckily for me I can pass and let my partner decide what to do; I must let him make the last mistake.'

5 PRIDAY: 'Well, that's an interesting bid of Claude's. He's obviously got a self-sufficient Heart suit and he's prepared to play in 4♡. Strictly speaking, I suppose I should leave him there, but it's interesting to look at what the defence is likely to be. They are likely to attack Diamonds and they won't know about my singleton if I'm not the dummy, so I think psychologically this hand could play better in Spades.'

6 GRANOVETTER: 'Thank you very much Mr Priday for letting me off the hook.'

7 RODRIGUE: 'Well, he doesn't want to play in Hearts. I shan't fight him.'

8 SILVERMAN: 'So what I've done to my partner, he's done back to

49

me. His pass is forcing so I have to do something. I don't want to bid my Diamonds at the five level when 5♣ could be a better spot. So I'll try 4NT which can't be Blackwood as I would have bid it the first time if I was interested in Aces, so Matt should take it for the minors.'

9 PRIDAY: 'I'm alright for the moment. I can leave the decision to Claude.'

10 GRANOVETTER: 'That must be the unusual No Trump. He probably has a hand like five Diamonds and three Clubs or six Diamonds and three Clubs. I'll just play it safe and rebid my club suit.'

11 RODRIGUE: 'I don't think anyone knows who's saving and who's expecting to make what on this hand. I'll leave it to Tony.'

12 SILVERMAN: 'I'm still not sure; but at least I know I have a fit.'

13 PRIDAY: 'Interesting. Matt has a lot of Clubs, Neil is supporting him, so Claude must be very short. It's difficult to know whose hand this is. Clearly we have two major suit winners in defence and we might make a second Heart or Spade. But generally on these sort of hands it is better to push on, so I'm not going to deceive opponents any more; this time I'll support partner.'

14 SILVERMAN: 'One day partner will make a decision for himself. I didn't know what to do before, and I certainly don't know what to do now. The one thing I do know is that I probably can't have a slam or my partner would have bid it himself. So the only question is, can they make 5♡ or not? It doesn't seem very likely, but I don't see how we can beat them more than one; so I'm just going to surprise my partner and pass.'

Well judged, as the contract went two down; and if the British women had gone one down as they should, then the Americans would have gained 7 imps on the board instead of losing a massive 15. Rodrigue was two down because he went all out for eleven tricks. The ♣A was cashed and the suit continued. Declarer ruffed and played a Spade to dummy. A Heart was led and finessed, North won and led a Diamond to South's Ace. Granovetter returned a Spade for North to ruff. The indulgent would describe Priday's 2♠ bid as conservative and the rest would call it an old woman's bid.

Board Sixteen, although not quite so expensive, was another horror story for the Americans, with poor defence in one room and a poor bid in the other.

Board Sixteen

Dealer East   Love All

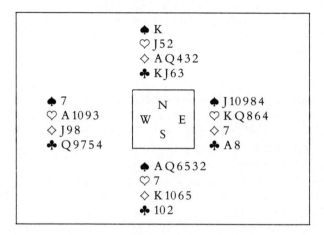

```
              ♠ K
              ♡ J52
              ◇ A Q 432
              ♣ K J 63
  ♠ 7                        ♠ J 10984
  ♡ A 1093      N           ♡ K Q 864
  ◇ J98       W   E         ◇ 7
  ♣ Q 9754      S           ♣ A 8
              ♠ A Q 6532
              ♡ 7
              ◇ K 1065
              ♣ 102
```

Room One

South	West	North	East
DAVIES, UK	MOSS, USA	GARDENER, UK	MITCHELL, USA
–	–	–	NB
1♠	NB	2◇	NB
3◇	NB	5◇	All pass

Contract: 5◇ by North. Lead: ♡Q.

After South had opened 1♠ and North had responded 2◇, South's second suit, the British women moved briskly to 5◇. Most inexperienced players would not open at all on the South hand; after all, it has only nine high-card points, but there are many times when points alone, or lack of them, are a delusion. South has six Spades, four Diamonds, only six losers, and is not vulnerable. Furthermore, the bid may prevent the opponents from getting together. A 1♠ opening in particular means that if they are going to compete, they are going to have to come in at the two level.

East led her ♡Q against 5◇, and Gail Moss went into a huddle before playing to the first trick: 'I know she's got the King so I could duck and let her have the problem as to what to do at trick two. However, if it is right to switch to a Club, she can't do it if she's looking at the ♣A and the ♣K is in Nicola's hand. I'm not sure which play is right; there are three choices, but Clubs is certainly one of them and therefore it

51

can't be wrong for me to win the lead.' Which she did, but was still left with the problem of what to lead to trick two: 'If Nicola has a doubleton Spade and my partner has the ◇A or another quick trick, perhaps I can get a Spade ruff and set the hand. That doesn't seem very likely; I think Nicola's Diamonds are probably very strong. I could play a trump, but that looks wrong. It seems as if the best chance looking at these menacing Spades in dummy is a Club. I hope partner has as good as the ♣AJ.'

So she led a Club which appears to give North an immediate problem, solved briskly and logically by Nicola Gardener: 'East led the ♡Q which means that West has the ♡A. Now then, if the ♣AQ are on my left (in the East hand) then I'm going down in this contract so I've got to assume they are split. Now since Gail has the ♡A, there's just a chance she's less likely to have the ♣A. Anyway, when two Aces are missing, I generally play for them to be divided. I'm going to play a low Club and let it run round to dummy's 10, hoping the ♣A is on my left.'

It is that sort of thinking that makes the difference between the good and the indifferent players. The poor player will go up with the ♣K almost without thinking and then bemoan his luck.

There are many bridge hands where, to make the contract, you have to assume a certain distribution and then play accordingly. If you were wrong, at least you were wrong for logical reasons and can claim to be unlucky. It is well worth spending a few moments on following Nicola Gardener's thinking here; it is a small point, and to the expert practically automatic, but not thinking as clearly as that costs the ordinary rubber bridge player hundreds of points a session.

The American defence was hardly distinguished. There are many Club holdings where to lead away from Queen to five will cost a trick. West should have led a second Heart to shorten dummy's trumps, give declarer entry problems and make it even more difficult to establish Spades, which was an unlikely line in any case. Sometimes one should play actively, sometimes passively; Gail Moss chose the wrong moment to get busy.

Room Two

South	West	North	East
GRANOVETTER, USA	RODRIGUE, UK	SILVERMAN, USA	PRIDAY, UK
–	–	–	NB
2♠	All pass		

Contract: 2♠ by South. Lead: ♣5.

Matthew Granovetter (South) decided to get busy too. His 2♠ open-ing bid was not only ill-judged, under the system he and Silverman were using it was quite simply wrong because his hand was far too *strong*. Normally, a weak Two bid shows no more than six to eight points and a reasonable six-card suit. His suit was all right but, he had nine extremely good points and a second suit. His smash-bang strategy not only silenced the opponents but his partner too. 2♠ presented no problem, but that was a bad result against the opponent's 5◇, and meant another 7 imps to the British putting them 41–28 in the lead.

Board Seventeen was a routine game contract, the Americans pulling back 1 imp thanks to Granovetter making an over-trick for a score of 650 against the British women's score of 620.

Board Seventeen

Dealer South   N-S Vulnerable

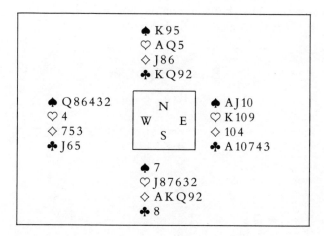

Room One

South	West	North	East
DAVIES, UK	MOSS, USA	GARDENER, UK	MITCHELL, USA
1♡	NB	2NT	NB
3◇	NB	3♡	NB
4♡	All pass		

Contract: 4♡ by South. Lead: ♠4.

Room Two

South	West	North	East
GRANOVETTER, USA	RODRIGUE, UK	SILVERMAN, USA	PRIDAY, UK
2♡¹	NB	2NT	NB
3◇²	NB	3♡³	NB
4♡	All pass		

Contract: 4♡ by South. Lead: ♣5.

1. Unabashed, he tries again. After a wretched experience last year, have the Americans up-graded their opening two bids?
2. Said to show a Diamond feature!
3. Not forcing. An invitation to game in Hearts leaving the decision to his partner.

If Granovetter's 2♠ on Board Sixteen was unwise, his 2♡ here is worse, but this time he gets away with it.

Board Eighteen was another swing to the British, this time of 5 imps. The hand showed the advantage of opening a major suit; by the time Gail Moss could show her suit, the bidding was already too high, and only eight tricks were made in both hands.

Board Eighteen

Dealer West   Game All

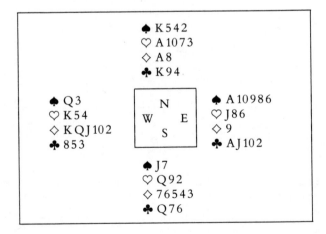

Room One

South	West	North	East
DAVIES, UK	MOSS, USA	GARDENER, UK	MITCHELL, USA
–	NB	1♡	1♠
2♡	3◇	All pass	

Contract: 3◇ by West. Lead: ♠2.

Room Two

South	West	North	East
GRANOVETTER, USA	RODRIGUE, UK	SILVERMAN, USA	PRIDAY, UK
–	NB	1♣[1]	1♠
NB	2◇	All pass	

Contract: 2◇ by West. Lead: ♠4.

1. The Americans play five card majors and a strong No Trump. 1♣ is therefore the only bid their system allows.

Board Nineteen

Dealer West   E-W Vulnerable

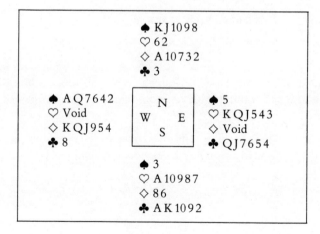

```
              ♠ K J 10 9 8
              ♡ 6 2
              ◇ A 10 7 3 2
              ♣ 3
♠ A Q 7 6 4 2      N         ♠ 5
♡ Void        W       E     ♡ K Q J 5 4 3
◇ K Q J 9 5 4     S         ◇ Void
♣ 8                         ♣ Q J 7 6 5 4
              ♠ 3
              ♡ A 10 9 8 7
              ◇ 8 6
              ♣ A K 10 9 2
```

55

Room One

South	West	North	East
DAVIES, UK	MOSS, USA	GARDENER, UK	MITCHELL, USA
–	1♠[1]	NB	2♡
NB	2♠[2]	NB	3♡
NB[3]	4♢[4]	Dble[5]	4♡[6]
Dble[7]	All pass		

Contract: 4♡ Doubled by East. Lead: ♣A.

1 MOSS: 'Oh, oh. This hand could be lots of fun or a nightmare.'

2 MOSS: 'It's beginning to look like a misfit. I'd like to show my second six-card suit, but I think I'd better make a non-forcing 2♠ bid and see if partner wants to go on.'

3 DAVIES: 'There's no way they are going to make 3♡ on this hand. I'm looking at three Heart tricks with an outside Ace. The problem is, if I double, have they somewhere better to go? I'm probably being a coward; I mean, I'm a notorious under-doubler but the thought of hearing them retreat to a better contract is just too painful to bear.'

4 MOSS: 'Can I never show this Diamond suit? I missed it last time, but how can I leave my partner in Hearts when we might have such a great fit in the Diamond suit I haven't bid yet? It feels wrong, but I'm going to bid my second suit.'

5 GARDENER: 'I've had enough of this auction.'

6 MITCHELL: 'Just stroll along to here.'

7 DAVIES: 'I can't believe it. I'll be able to tell everyone who accuses me of not doubling how I gained an extra 300 by not doubling in the first place so that I could double later on. This surely must be going for a packet.'

And so it did, East making only five tricks for a crunching penalty of 1,400 to the British. Meanwhile, the British men had to play those same E-W cards.

Room Two

South	West	North	East
GRANOVETTER, USA	RODRIGUE, UK	SILVERMAN, USA	PRIDAY, UK
–	1♠	NB^1	2♡
NB^2	3◇^{3}	NB^4	3♡^{5}
NB	4◇^{6}	Dble^7	4♠^{8}
Dble^9	All pass		

Contract: 4♠ Doubled by West. Lead: ♣3.

1 SILVERMAN: 'Hey, I was going to bid one of those.'

2 GRANOVETTER: 'What is this? I thought I was going to be in the bidding. I could bid 3♣ but just maybe they have a misfit. . . .'

3 RODRIGUE: 'I should have more high card points to bid 3◇, but really the shape justifies it.'

4 SILVERMAN: 'When this hand is over, I have the feeling Claude is going to wish he was my partner.'

5 PRIDAY: 'I was afraid this might turn out to be a fairly substantial misfit, but I can't do any more than just repeat my suit.'

6 RODRIGUE: 'My hand doesn't look any better, but I can't pass for two reasons. The first is that 3♡ is forcing and the other is that I've got 12 cards in Spades and Diamonds and so far I've only shown a four-card Diamond suit. I suppose I'd better see it through, but I really am beginning to dislike this hand.'

7 SILVERMAN: 'I don't honestly think they're going to play in 4◇, but now's the time to let partner know that things are well placed for us.'

8 PRIDAY: 'Well, it had to come. I'm not very hopeful of 4♠, but at least I do have one.'

9 GRANOVETTER: 'I'm not taking any chances on this one. I have Ace-King, Ace and my partner doubled. I'm going to sock this.'

In fact, the British men had found a marginally less horrible resting place than the American women, and even though 4♠ Doubled went four down for a penalty of 1,100, the net gain on the board was 300, or 7 imps to the British.

After that short-arm jab to the solar plexus, came this right cross; Board Twenty.

Board Twenty

Dealer North   Love All

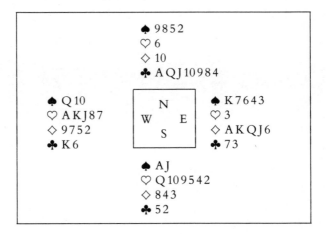

♠ 9852
♡ 6
♢ 10
♣ A Q J 10 9 8 4

♠ Q 10
♡ A K J 8 7
♢ 9 7 5 2
♣ K 6

♠ K 7 6 4 3
♡ 3
♢ A K Q J 6
♣ 7 3

♠ A J
♡ Q 10 9 5 4 2
♢ 8 4 3
♣ 5 2

Room One

South	West	North	East
DAVIES, UK	MOSS, USA	GARDENER, UK	MITCHELL, USA
–	–	3♣	3♠
NB	3NT	All pass	

Contract: 3NT by West. Lead: ♡6.

The bidding in Room One was straightforward, but the play certainly was not. One is always told to lead the fourth highest of your longest suit against a No Trump contract or, as with the North hand, the highest of an internal sequence, in this case the ♣Q. Nicola Gardener had different ideas.

'Well, there's a five-card Spade suit on my left and Gail on my right has bid 3NT so the one thing that is certain is that she has the ♣K. If I lead a Club, I am giving up a Club trick, but maybe I should do so in order to make the rest. On the other hand, there must be a chance that if I can get Pat in, then she can lead a Club through and we can make them all. So what the hell am I going to lead? I don't want to lead a Spade; that's Jackie's suit. A Heart or a Diamond – a singleton against 3NT doesn't look right, but it has to be one of the two. Which Ace has partner got? If she hasn't got one, I'm not going to beat this contract, so let's think positively and lead a Heart and see what happens.'

South went up with the Queen, declarer won with the Ace and led a Diamond to the table, on which North played her singleton 10. West has a real problem on this hand on that Heart opening lead. She has just eight tricks, five Diamonds and three Hearts; the difficulty is to find the ninth without letting South in to lead a Club through. Obviously, West cannot lead a Club from dummy herself. So, her only option is to lead a Spade and hope either North has the Ace in which case nine tricks are certain or that South has the Ace but ducks for some reason. So at trick three, West led a small Spade from dummy which gave South cause for thought.

Pat Davies: 'Well, there can't be much percentage in ducking this Spade; its just possible declarer has nine tricks if I do duck. So I had better go up with the Ace, but the question is; what to do next?' (Whereupon Pat Davies went into a long trance while her partner managed stoically but under intense duress, to remain totally expressionless as demanded by the ethics of Contract Bridge.) 'Under our system, the ♡6 was either a singleton or she has three to an honour. Now, when Diamonds were played she went up with the ten. If she plays high-low in Diamonds, then she wants me to return the suit she led. So, either she wants me to play back a Heart or she had a singleton Diamond in which case she had no choice but to play the ten. Can she really have a singleton Diamond and a singleton Heart? Jackie overcalled 3♠ and Gail didn't support her, so presumably Gail's only got a doubleton in which case, Nicola has four Spades. She opened 3♣, so presumably she has seven Clubs. That seems right. She has a singleton in both red suits and she was forced to play a high Diamond; she didn't play it in order to encourage me to play the Heart back. Oh well, I've made her sweat a bit while I've checked all this about the distribution, but I'm pretty sure now that I know what's right.'

After this thoroughly commendable analysis she led a Club, West went up with the King and that was the end of the story – four down as Nicola gratefully cashed her seven Club winners.

Room Two

South	West	North	East
GRANOVETTER, USA	RODRIGUE, UK	SILVERMAN, USA	PRIDAY, UK
–	–	NB[1]	1♠
NB	2♡	3♣[2]	NB
NB	Dble[3]	NB	3◇[4]
NB	3NT[5]	All pass	

Contract: 3NT by West. Lead: ♣Q.

1 SILVERMAN: 'Quite a nice Club suit. Unfortunately, in our style, this hand's a little too big for 3♣. Besides, I have a four-card Spade suit on the side which could conceivably play in 4♠ if partner had something like four to the KQ. So, If I open 3♣ I could pre-empt our side.'

2 SILVERMAN: 'The time has come to bid my suit. There's no longer the threat that we're going to play in 4♠ and I want to make sure my partner leads the right suit if he happens to get the lead.'

3 RODRIGUE: 'Well, Neil seems to have come out of the undergrowth now. The question is – are we going to step on him? Tony hasn't got more than two Hearts; if he had, he'd be supporting me, and we've certainly got the balance of the high cards. If I double, its a co-operative sort of double and if Tony doesn't like it, he can take it out.'

4 PRIDAY: 'The trouble is, partner doesn't know my main strength is in Diamonds. I have this solid suit, he may have a few himself and it could be we won't be able to take enough tricks to make the penalty worth while. At least I can show my Diamonds without showing any extra strength.'

5 RODRIGUE: 'Its all rather frustrating. Tony seems rather distributional. Maybe I'll be able to run nine tricks and they won't be able to get at me in No Trumps quickly enough.'

Neil Silverman did lead a Club, but not before a little thought. 'This is the problem with my Clubs being a little too good. I could try to lead a side suit and hope my partner gets in to play a Club through, but the likelihood is that Claude has three Clubs to the King. It seems much more likely that partner has the Diamond or Spade suit stopped and Claude might not be able to come to nine quick tricks.'

Claude Rodrigue won the Club lead and cashed his five Diamond winners: 'Now I have two top Hearts which makes eight and I could try and set up a Spade trick with the combined K and Q but what have opponents been discarding? Neil on my left has discarded the ♣A to show his Clubs are solid, and apart from that, he's discarded a Club, a Heart and a Spade while Matthew hasn't discarded any Clubs at all. I think I can infer that Matthew started with a doubleton Club and Neil is sitting waiting to cash four Club tricks as soon as his side gets the lead. Whoever has the ♠A is going to go up with it and the Clubs are going to be cashed. So, I can't set up a Spade trick and I'm going to have to take the Heart finesse.'

A small Heart from dummy, South played low, Rodrigue inserted the ♡J. 3NT made and another 12 imps to the British.

Although Nicola's lead in Room One worked well, many experts would consider it against the odds.

Board Twenty-one was level, both E-Ws bidding and making 6♠. What was slightly odd was that neither side bid to the grand slam, although if they had, they would not have made it, thanks to the trump distribution, unless East had taken a double-dummy view of the Spades and finessed the ten on the second round. A trump reduction does not succeed because declarer cannot cash his Club winners.

Board Twenty-one

Dealer East   N-S Vulnerable

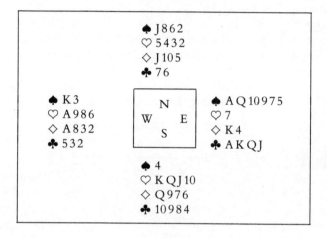

	♠ J862	
	♡ 5432	
	◇ J105	
	♣ 76	
♠ K3	N	♠ A Q 10 9 7 5
♡ A986	W E	♡ 7
◇ A832	S	◇ K4
♣ 532		♣ A K Q J
	♠ 4	
	♡ K Q J 10	
	◇ Q976	
	♣ 10984	

Room One

South	West	North	East
DAVIES, UK	MOSS, USA	GARDENER, UK	MITCHELL, USA
–	–	–	1♠
NB	2NT	NB	3♣
NB	3♡	NB	3♠
NB	4♠	NB	4NT
NB	5♡	NB	6♠
All pass			

Contract: 6♠ by East. Lead: ♡K.

Room Two

South	West	North	East
GRANOVETTER, USA	RODRIGUE, UK	SILVERMAN, USA	PRIDAY, UK
–	–	–	2♣
NB	2NT	NB	3♣
NB	4◇	NB	4♠
NB	5♡	NB	6♣
NB	6♠	All pass	

Contract: 6♠ by East. Lead: ♡K.

But having had this board to get their breath back, the Americans now suffered another thudding jolt on Board Twenty-two where Jackie Mitchell at trick three had a chance to shine but came up with the wrong answer to give another 10 imps to the United Kingdom.

Board Twenty-two

Dealer South   E-W Vulnerable

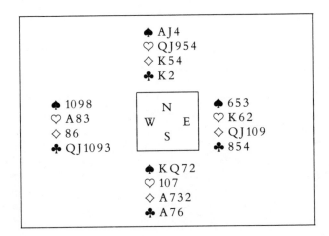

```
                ♠ A J 4
                ♡ Q J 9 5 4
                ◇ K 5 4
                ♣ K 2
   ♠ 1098            N           ♠ 653
   ♡ A 8 3       W       E       ♡ K 6 2
   ◇ 86              S           ◇ Q J 10 9
   ♣ Q J 10 9 3                  ♣ 8 5 4
                ♠ K Q 7 2
                ♡ 10 7
                ◇ A 7 3 2
                ♣ A 7 6
```

Room One

South	West	North	East
DAVIES, UK	MOSS, USA	GARDENER, UK	MITCHELL, USA
1NT	NB	2◇	NB
2♡	NB	3NT	All pass

Contract: 3NT by South. Lead: ♣Q.

Room Two

South	West	North	East
GRANOVETTER, USA	RODRIGUE, UK	SILVERMAN, USA	PRIDAY, UK
1◇	NB	1♡	NB
1♠	NB	2♣	NB
2NT	NB	3NT	All pass

Contract: 3NT by South. Lead: ♣Q.

There was nothing of particular interest in the bidding in either room, except that the British women's predeliction for the weak No Trump allowed them to reach 3NT without bidding every suit in the pack first.

The play, however, was instructive. In Room One, South ducked the Club lead, then perforce won the continuation on the table. Now, she always has eight tricks on top: four Spades, two Diamonds and two Clubs. If she can establish a Heart before the opponents can establish Clubs, then there is her ninth trick. But, if West has a Heart honour and East can knock out South's ♣A before West has to use her Heart entry, then the defence will come to two Hearts and three Clubs to defeat the contract. So, at trick three, Pat Davies put Jackie Mitchell, the American East, firmly on the spot by leading a low Heart from the table. She calculated wrongly, played low, West won with the Ace, but now had no entry to enjoy her long Club and so had the mortification of seeing South stroll home in 3NT.

In Room Two, the American South made the British defence much easier by cashing his Spades first and then, after a long trance, leading a Heart towards dummy which West had no hesitation in ducking. East won with the ♡K, shot back a Club, and West still had the ♡A as an entry for his Clubs to defeat the contract by one trick and gain 10 imps for the United Kingdom.

To be fair to Granovetter he had to decide between the technical chance of the 3–3 Diamond break, or inducing the opponents into error by playing a Heart.

Board Twenty-three was played quietly in 2♠ in each room, both Norths making an over-trick.

Board Twenty-three

Dealer West   Game All

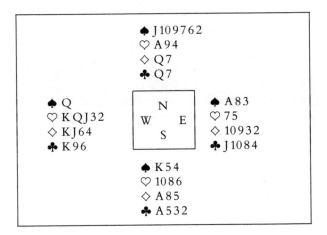

	♠ J109762	
	♡ A94	
	◊ Q7	
	♣ Q7	

♠ Q	N	♠ A83
♡ KQJ32	W E	♡ 75
◊ KJ64	S	◊ 10932
♣ K96		♣ J1084

	♠ K54	
	♡ 1086	
	◊ A85	
	♣ A532	

Room One

South	West	North	East
DAVIES, UK	MOSS, USA	GARDENER, UK	MITCHELL, USA
–	1♡	1♠	NB
2♡	Dble	2♠	All pass

Contract: 2♠ by North. Lead: ♡7.

Room Two

South	West	North	East
GRANOVETTER, USA	RODRIGUE, UK	SILVERMAN, USA	PRIDAY, UK
–	1♡	1♠	NB
2♡	NB	2♠	All pass

Contract: 2♠ by North. Lead: ♡7.

Board Twenty-four was a delight for all who like to see experts suffer. At first it looked as if the British women, who went on down, had sacrificed quite wrongly and that the board was going to show a considerable loss for the United Kingdom. But the tide was still running full against the Americans and they managed to turn victory into defeat.

Board Twenty-four

Dealer North   N-S Vulnerable

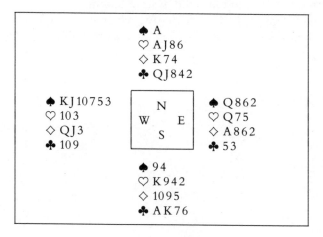

♠ A
♡ A J 8 6
◇ K 7 4
♣ Q J 8 4 2

♠ K J 10 7 5 3   ♠ Q 8 6 2
♡ 10 3   ♡ Q 7 5
◇ Q J 3   ◇ A 8 6 2
♣ 10 9   ♣ 5 3

♠ 9 4
♡ K 9 4 2
◇ 10 9 5
♣ A K 7 6

Room One

South	West	North	East
DAVIES, UK	MOSS, USA	GARDENER, UK	MITCHELL, USA
–	–	1♣	NB
1♡	2♠	4♡	4♠
5♣	All pass		

Contract: 5♣ by North. Lead: ♠2.

Room Two

South	West	North	East
GRANOVETTER, USA	RODRIGUE, UK	SILVERMAN, USA	PRIDAY, UK
–	–	1♣	NB
1♡	1♠	3♡	3♠
4♣	4♠	NB	NB
5♣	NB	6♣	NB
6♡	All pass		

Contract: 6♡ by South. Lead: ♠J.

Six Hearts is a preposterous contract; not a good advertisement for an American systemic gadget which we lack the space to explain fully. It will go one down, probably two, against best defence, and in practice even without a Diamond lead did go two down.

Another 3 imps to the British, an overwhelming session in which the United Kingdom gained no less than 56 imps to lead by 78–29.

Session Three

BOARDS TWENTY-FIVE TO THIRTY-SIX
'Thanks for the memory, and a little help from my friends.'

A year ago, a British player went five off doubled in a grand slam missing an Ace. Haunted by the memory, she eschewed the grand slam on Board Twenty-seven while on an up-beat of abandoned enthusiasm, Granovetter not only punted the grand slam but had the technique to make it.

For a couple of boards it seemed the Americans had stemmed, if not turned, the tide, but then on Board Twenty-nine Granovetter, seized with doubts, settled for a game while Gardener and Davies were pugnaciously bidding a small slam to gain 11 imps. Ten of which were instantly lost on the quirkiness of Board Thirty.

Humpty Dumpty made words mean what he wanted them to mean; some Bridge players do the same with their bids and then castigate their partners for not being mind readers. The American women are too disciplined to do that, and Gail Moss chose to lose a few imps rather than violate principles and possibly cause her partner to distrust her for the rest of the match.

It's interesting to compare her reasoning on Board Thirty-four with the way Rodrigue once or twice in the match revalued his hand slightly to make bidding easier. Bridge is far too imprecise for the slavish addiction to dogma.

Board Twenty-five

Dealer North   N-S Vulnerable

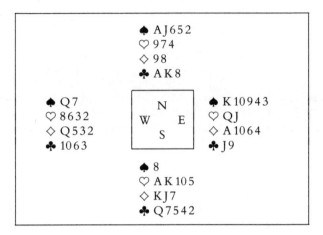

```
                    ♠ A J 6 5 2
                    ♡ 9 7 4
                    ◇ 9 8
                    ♣ A K 8
    ♠ Q 7          ┌─────────┐      ♠ K 10 9 4 3
    ♡ 8 6 3 2      │    N    │      ♡ Q J
    ◇ Q 5 3 2      │ W     E │      ◇ A 10 6 4
    ♣ 1 0 6 3      │    S    │      ♣ J 9
                    └─────────┘
                    ♠ 8
                    ♡ A K 10 5
                    ◇ K J 7
                    ♣ Q 7 5 4 2
```

Room One

South	West	North	East
DAVIES, UK	MOSS, USA	GARDENER, UK	MITCHELL, USA
–	–	1♠	NB
2♣	NB	3♣	NB
3NT	All pass		

Contract: 3NT by South. Lead: ◇2.

Room Two

South	West	North	East
GRANOVETTER, USA	RODRIGUE, UK	SILVERMAN, USA	PRIDAY, UK
–	–	1♠	NB
2♣	NB	3♣	NB
3NT	All pass		

Contract: 3NT by South. Lead: ♡6.

The British made two overtricks, the Americans in Room Two, only one, which was enough to give another imp to the United Kingdom.

Board Twenty-six was equally straightforward, only this time it was the Americans who sneaked an overtrick and with it an imp.

Board Twenty-six

Dealer East   E-W Vulnerable

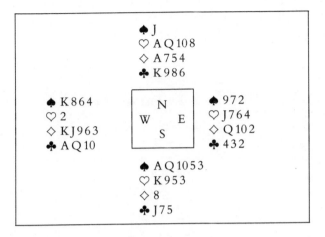

♠ J
♡ A Q 10 8
♢ A 7 5 4
♣ K 9 8 6

♠ K 8 6 4
♡ 2
♢ K J 9 6 3
♣ A Q 10

N
W   E
S

♠ 9 7 2
♡ J 7 6 4
♢ Q 10 2
♣ 4 3 2

♠ A Q 10 5 3
♡ K 9 5 3
♢ 8
♣ J 7 5

Room One

South	West	North	East
DAVIES, UK	MOSS, USA	GARDENER, UK	MITCHELL, USA
–	–	–	NB
NB	1◇	1♡	NB
2◇	NB	3◇	NB
4♡	All pass		

Contract: 4♡ by North. Lead: ◇2.

Room Two

South	West	North	East
GRANOVETTER, USA	RODRIGUE, UK	SILVERMAN, USA	PRIDAY, UK
–	–	–	NB
NB	1◇	1♡	NB
4♡	All pass		

Contract: 4♡ by North. Lead: ◇2.

Both declarers opted for a cross-ruff; Nicola Gardener came to ten tricks, Neil Silverman to eleven to gain one imp. It is interesting, by the way, to notice that it is important to ruff the Spades before the Diamonds. On the bidding, it is always possible that East, who is short in Diamonds, will be able to discard a Spade which will leave declarer one trick short.

Board Twenty-seven was the first genuine grand slam of the match; and the British women correctly missed it, to give the Americans their biggest swing so far, perhaps helping them to feel that Lady Luck was not, after all, British.

Board Twenty-seven

Dealer South   Game All

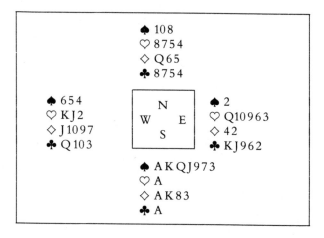

Room One

South	West	North	East
DAVIES, UK	MOSS, USA	GARDENER, UK	MITCHELL, USA
2♣[1]	NB	2◇[2]	NB
3♠[3]	NB	4♠[4]	NB
5◇[5]	NB	6♠[6]	NB
NB[7]	NB		

Contract: 6♠ by South. Lead: ◇J.

1 Unconditionally forcing.

2 Conventional, negative bid.

3 DAVIES: 'I could show my suit by simply bidding 2♠ here, but

Nicola and I play a jump to 3♠ showing a solid suit, setting the trumps and asking partner to cue bid. Then I can find out if she has any useful cards for me.'

4 GARDENER: 'Well, Pat's got game in her own hand. I've nothing much to add, but I'll put her in game.'

If Nicola had bid 3NT, denying an Ace, it would have left Pat Davies room to find out about the all important third round Diamond control. The bidding might then have continued: N: 3NT, S:4◇, N:4♠ (denying second round controls), S:5◇, N:7♠. But who wants to be in a grand slam which has less than a 50% chance of success?

5 DAVIES: 'That's the weakest bid Nicola can make. If she had any sort of control, she would have shown it, so she can't have the ♡K or ♣K. I've got a big enough hand to look further; the question is, what is the best way to look? I could bid 5♣ showing a Club control, but on this sort of hand, we usually bid length, not a suit where we're short, and that helps partner to judge whether her cards are working or not. So, if I bid Diamonds it'll ask partner for help in Diamonds.'

6 GARDENER: 'That's improved my hand a little. Pat is likely to have at least three Diamonds, and she could get into my hand with the ♠10 since I know she has the AKQJ. I really like the look of this ◇Q; and I'm not known for under-bidding.'

7 DAVIES: 'So Nicola's got some help in Diamonds, probably the Queen. The question is the Queen to how many? If she has three Diamonds to the Queen, I've got the problem of what to do with the fourth Diamond. I know she hasn't the ♡K or ♣K on which to discard my fourth Diamond and I can't be certain of ruffing because she doesn't promise any Spades by the raise to 4♠. On the other hand, she could have the ◇Q doubleton and two trumps, when the grand slam is on. This is really a guessing game. The trouble is if you bid a grand slam and it goes off, that's a swing of 17 imps and I haven't forgotten that I achieved infamy in the last series as the person who played 7NT doubled and lost 1100. I'm fighting a bit shy of grand slams in this match.'

So she passed and playing for safety, made precisely 12 tricks and her contract.

Room Two

South	West	North	East
GRANOVETTER, USA	RODRIGUE, UK	SILVERMAN, USA	PRIDAY, UK
2♣	NB	2◇	NB
2♠[1]	NB	3♣[2]	NB
3◇[3]	NB	4♠[4]	NB
7♠[5]	All pass		

Contract: 7♠ by South. Lead: ♠5.

1 GRANOVETTER: 'No matter what, I'm going to a slam on this hand, but I'm going to try to get some information out of Neil. I'll continue with 2♠ which is forcing; at least, he had better not pass.'

2 SILVERMAN: 'Well, I get the chance to show a really awful hand. 3♣ in our system denies even a King.'

3 GRANOVETTER: 'I really have an interesting choice here. To bid my Diamond suit to try to get information out of Neil – perhaps we can make 7♠ if I can get a raise in Diamonds – or to conceal the suit which would be most advantageous in the play because the opponents will have a very difficult time discarding while I'm running my Spades if they don't know I have the Diamonds. But I must get some information. It's a team game and I want to be in Seven if Seven is there.'

4 SILVERMAN: 'Well, now I finally like my hand. It's hardly inspiring with only one Queen, but that Queen is in Matthew's second suit and I do have two trumps in his original suit, so I'd better come to life here.'

5 GRANOVETTER: '*Four* Spades. He must have liked something about my bidding. The only thing I've shown him is Spades and Diamonds. Well, he doesn't have much in Spades. He must have a singleton or doubleton Diamond, maybe the Queen doubleton and three trumps. Well, I've got something out of Neil and I'm not going to be timid now.'

Granovetter won the Spade lead in dummy, attacked Diamonds straight away and when East discarded on the third round, claimed his contract by means of a ruff of the fourth Diamond; that was the ♣A, ♡A, three top Diamonds, a Diamond ruff and seven Spade tricks: plus 13 imps to the USA. It is very interesting but Seven Spades virtually depends upon a 3–3 Diamond break. This is about a 36% chance and therefore poor odds for a grand slam. Granovetter, however, took advantage of the additional chance of the Diamonds breaking 4–2 with the short Diamonds with the singleton trump, as was the case. Rodrigue might have deceptively followed with the ◇10 on the Ace and continued with the ◇J on the next round. Declarer might then have decided to try to draw the trumps before finessing the ◇8 and plunging to his doom. We shall never know!

Board Twenty-eight was charming; a moment of anguish for each of the American men, and in the end the delight of 4 imps well earned.

Board Twenty-eight

Dealer West   Love All

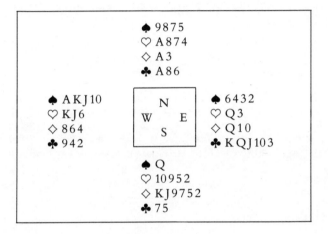

Room One

South	West	North	East
DAVIES, UK	MOSS, USA	GARDENER, UK	MITCHELL, USA
–	1♣[1]	Dble[2]	Redble[3]
2◇[4]	NB[5]	NB	3♣
All pass			

Contract: 3♣ by West. Lead: ◇A.

1 MOSS: 'Well, it would be generous to say this is an optional opening bid. Twelve high card points, no distribution, but I do have good cards in the majors and two and a half quick tricks.'

2 GARDENER: 'I'm playing with a particularly timid partner who needs

73

every encouragement. I may be weak, but after all, I've two four-card majors.'

3 MITCHELL: 'Due to the fact that I have ten points in high cards, well sort of, I guess I'm going to redouble.'

4 DAVIES: 'I could just bid 1◇, but I might as well get over the fact I've got a six-card suit. 2◇ wouldn't show any values; I'm simply saying I have six Diamonds.'

5 MOSS: 'Glad to hear Jackie's so enthusiastic about this hand. I'm certainly not. I wasn't to start with, and Nicola sitting on my left with the majors has killed it, what little there was of it.'

6 MITCHELL: 'Well, I don't know, but it seems some sort of right to support Clubs here.'

After the lead of the ◇A and a small Club by North, Gail Moss duly harvested the nine tricks she needed to make 3♣.

Room Two

South	West	North	East
GRANOVETTER, USA	RODRIGUE, UK	SILVERMAN, USA	PRIDAY, UK
–	1NT[1]	NB	2♣[2]
2◇[3]	2♠[4]	2NT[5]	NB[6]
3◇[7]	NB	NB[8]	3♠
All pass			

Contract: 3♠ by West. Lead: ◇A.

1 RODRIGUE: 'The only possible bid on this hand the way we play is 1NT. Its a very yuk No Trump, but still I'll bid it.'

2 PRIDAY: 'I rather like my hand in that it's got a solid Club suit that is certainly going to make tricks and has the possibility of a Spade fit. I can certainly look for that safely enough by a Stayman bid.'

3 GRANOVETTER: 'I don't know. I think I'll make a little, biddy lead director here.'

4 RODRIGUE: 'I'm off the hook, if I want to be, but unless Tony is mucking around, my hand has improved. My four-card major really is superb.'

5 SILVERMAN: 'There's an awful lot of bidding around. And this could easily be our hand. I'll just bid 2NT and see what partner does.'

6 PRIDAY: 'What's going on here? Claude has shown a No Trump opening, I've got ten points and they're bidding 2NT. Someone is being funny at the table, and I don't think it's us. Perhaps I should support Claude, but my Spades are very weak so I think I'll just bide my time.'

7 GRANOVETTER: 'No partner, dear. It was not an invitation to play in 3NT. It was a lead director, that's all; a little lead director.'

8 SILVERMAN: 'Well, they always say it's right to trust your opponent and not your partner.'

Although the Americans may appear venturesome nine tricks can in fact be made in either red suit. Against West's actual 3♠, Neil Silverman led ◇A and another Diamond won by South's King. And while Granovetter went into one of those prolonged trances that give the other players time to catch up with their mail, do a couple of crosswords or otherwise usefully occupy their time, Silverman had plenty of time for reflection. 'As usual, Matthew's thinking and the more he's thinking, the more I think I made a mistake. (Any kitchen Bridge partnership would have broken this contract at once. North would have led his three Aces, a Diamond to South's King and ruffed the Diamond return for the first five tricks. Sometimes the experts go on a little tour to arrive at the same result.) I could certainly have cashed my other two Aces before playing the Diamond, then he would have had to play a Diamond to set up my low Spade. Even if he doesn't play a Diamond now, I don't think Claude will be able quite to handle this hand. Still, I hope he plays a Diamond.'

But Matthew Granovetter, who has the soul of an artist, led not a Diamond but a Heart, and in the end had the subtle pleasure of promoting one of his partner's trumps for the setting trick instead of the crude joy of a smash bang Diamond ruff. So that the reader can enjoy the artistry, here was what happened in detail. Neil Silverman won the Heart return with his Ace and returned a Heart to kill dummy's quick entry. Declarer played a Spade and noting South's Queen, switched to Clubs. North, however, was in full control of the situation and ducked the first Club, won the second, and returned a Club, killing dummy's entry yet again and leaving declarer with an inescapable fifth loser, either a trump or a Diamond. 3♠ one down and 4 imps to the Americans.

Board Twenty-nine; pugnacious accuracy from the British women, exaggerated pessimism from the American men, and another handsome swing to the British.

75

Board Twenty-nine

Dealer North   E-W Vulnerable

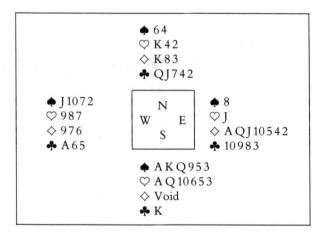

```
              ♠ 64
              ♡ K 4 2
              ◇ K 8 3
              ♣ Q J 7 4 2
  ♠ J 10 7 2      N        ♠ 8
  ♡ 987      W       E     ♡ J
  ◇ 976                    ◇ A Q J 10 5 4 2
  ♣ A 6 5         S        ♣ 10 9 8 3
              ♠ A K Q 9 5 3
              ♡ A Q 10 6 5 3
              ◇ Void
              ♣ K
```

Room One

South	West	North	East
DAVIES, UK	MOSS, USA	GARDENER, UK	MITCHELL, USA
–	–	NB	3◇
4◇[1]	NB	6♣[2]	NB
6◇[3]	NB	6♡	All pass

Contract: 6♡ by North. Lead: ◇A.

1 DAVIES: 'Usually we play 4♣ for take out, but that would show all the other three suits. 4◇ shows a big, two suited hand – which is exactly what I have.'

2 GARDENER: 'So Pat has an enormous two-suiter. It looks like Spades and Hearts but it could equally well be Clubs and Spades or Clubs and Hearts. My hand's terrible in Spades, and not much better, in Hearts. I could bid 5◇ asking Pat to choose, but I don't want to do that; it's such a nebulous bid because even if Pat bids 5♡, Clubs could still play better and I shan't know which two suits she has even then. Of course, if I jump to a slam in Clubs, perhaps Pat will get the message that I'm weak in another suit. I can't have a solid seven-card Club suit (*Gardener passed on the first round of bidding*), and she's not going to think that, is she? No, she wouldn't do that to me.'

3 DAVIES: 'Nicola passed originally, so her 6♣ doesn't say she wants to play in Clubs regardless of what I've got. All it means is that if one of my suits is Clubs, she's happy to play there and if I've got Hearts or Spades, she's happy to be there. 6◇ will ask her to choose.' Which she does.

Gardener ruffed the ◇A in dummy, played a Heart, to the King, discarded the ♣K on the ◇K, played a Spade to the Ace, drew a second round of trumps and then established the Spades to make all 13 tricks.

Room Two

South	West	North	East
GRANOVETTER, USA	RODRIGUE, UK	SILVERMAN, USA	PRIDAY, UK
–	–	NB	3♣[1]
3◇[2]	NB	3♡[3]	NB
4◇[4]	NB	4♡[5]	NB
NB[6]	NB		

Contract: 4♡ by North. Lead: ♠8.

1 A transfer pre-empt, allowing partner to play the hand.

2 GRANOVETTER: 'I'm glad they're using this transfer because now I can simply show both majors.'

3 SILVERMAN: 'So he's got a good hand with at least nine or ten cards in the majors. It's possible I might be able to make 3NT since I do have the Diamond stopper. But I sort of have just a bunch of junk; it'd be difficult to make nine tricks. I think I'll just pick my better major.'

4 GRANOVETTER: 'Well I'm certainly bidding to game; its just a question of whether we have a slam. The only slam try I have is 4◇ which should say to partner, if you have something there, help me out.'

5 SILVERMAN: 'I don't think any of my minor suit values are going to be of help to him as more than likely he has a void Diamond so the King is wasted. The ♣QJ may help a little, but I don't think the ♡K is quite enough to go to the five level; I think I'll just wait and hope he makes one more try.'

6 GRANOVETTER: 'Oh crap. What a position. Shall I continue or not? I forced Neil to bid a major. If he had been 2–2, he would have had to bid a major. Then I forced him to repeat it. He could have three Hearts, but then, that was a vulnerable pre-empt by Tony. Things could split badly. I need just the right cards for a slam. I'm going to take my plus score.'

77

It was the sign of an excellent player to realise his partner might have only two Hearts, but an also interesting example of the pessimistic introversion that sometimes overwhelms Granovetter that he finally convinced himself partner did indeed have only two.

Board Thirty

Dealer East   E-W Vulnerable

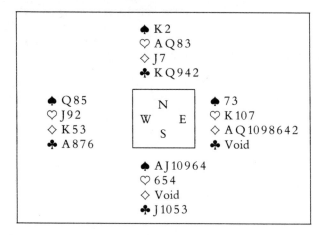

Room One

South	West	North	East
DAVIES, UK	MOSS, USA	GARDENER, UK	MITCHELL, USA
–	–	–	3◇[1]
NB	3NT[2]	NB	5◇[3]
All pass			

Contract: 5◇ by East. Lead: ♣J.

Room Two

South	West	North	East
GRANOVETTER, USA	RODRIGUE, UK	SILVERMAN, USA	PRIDAY, UK
–	–	–	NB
3♠	NB	NB	4◇
All pass			

Contract: 4◇ by East. Lead: ♣3.

1 MITCHELL: 'Nothing I like better than eight-card suits. I never have the slightest idea what to do with them. I'd like to open 5◊, but vulnerable against non-vulnerable, I'd best start with 3◊ and hope I don't miss something.'

2 MOSS: 'What a peculiar hand. when your partner pre-empts and you don't have any great distribution and certainly not an inordinate number of points, you don't think in terms of game. Yet the minimum my partner could have in Diamonds would be AQ seven times and if she's as good as the Jack and another Spade and the Queen and another Heart, we easily might be able to take nine tricks in no time. Just as important, it might not even be our hand; the opponents might be in a position to make game here. I think the best thing to do is bid 3NT. If we get set, it might very well turn out it was opponents' hand for a game. I might just intimidate them.'

3 MITCHELL: 'Oh, oh; back where I started. She'll probably kill me if I bid. Nothing like pre-empting and then finding another bid when partner is not in the least interested. I wanted to open this hand 5◊, then I chickened out, but now that Gail has some cards I think far more often than not that I'll be safer in 5◊ than she will be in 3NT. I'll bid 5◊ and never look across the table at her face.'

Curiously, with North on lead, 3NT is stone cold as the cards lie, while on any lead but a Club, 5◊ is almost certainly going down. However, South (Pat Davies) reasoned that with her holding, if East had a void it was likely to be in Spades – and led a Club. With the ♡AQ both in North's hand, on a Club lead 5◊ was virtually a lay down.

In Room Two, the Americans found the same defence against the superior contract of 4◊, so the British made an over-trick but lost 450 on the board which was 10 imps to the Americans.

Who says Bridge is a game of pure skill?

So, on these six boards, the USA pulled back 16 imps to trail 57–90.

Board Thirty-one

Dealer East   Game All

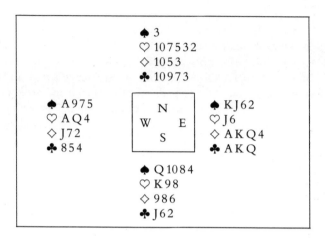

Room One

South	West	North	East
DAVIES, UK	MOSS, USA	GARDENER, UK	MITCHELL, USA
–	–	–	2♣[1]
NB	2◇[2]	NB	2NT[3]
NB	6NT[4]	All pass	

Contract: 6NT by East. Lead: ♠4.

Room Two

South	West	North	East
GRANOVETTER, USA	RODRIGUE, UK	SILVERMAN, USA	PRIDAY, UK
–	–	–	2♣
NB	2NT	NB	3NT
NB	4♣	NB	4◇
NB	4♠	NB	6♠
All pass			

Contract: 6♠ by West. Lead: ♡2.

1 MITCHELL: 'It's a No Trump hand, but too good to open 2NT.'

2 MOSS: 'What a surprise. Partner opens with the strongest bid on our system and I've got a beautiful hand myself. If I were to bid a suit, that would guarantee at least a five-card suit, and I don't have one. I could give a positive response in No Trumps, but the problem with that is that I suspect this hand is very likely to end in some number of No Trumps, probably a slam, and with my Diamond and Club holdings, it would be much better to play from my partner's side where her honours will be protected. That just leaves 2◇ which theoretically is negative but can also be a waiting bid. I'll compensate for the weakness of the bid later.'

3 MITCHELL: 'If partner really does have a negative, then she can pass 2NT.'

4 MOSS: 'Delighted I made the choice I did. Now the hand is playing from the right side of the table. She has at least 22 high card points; we must have 33 between us so it's impossible for opponents to cash two Aces.'

South led the ♣4, reasoning that it was the sort of hand where the defenders needed to be active. But as the bidding suggested that the bulk of the strength was on her right, it was a bad lead in theory and in practice. This was the sort of hand where passive defence might have left an average declarer a chance to go wrong. Against Jackie Mitchell it would not have helped. She would win a minor suit lead in hand and immediately test Hearts by finessing the Jack. South would cover but declarer then has eleven top tricks. The twelfth is guaranteed by cashing the ♣K and then leading the ♣2 and covering South's card. Should South show out, the Ace is played and a low spade led through North's Queen.

In Room Two, West received the distinctly friendly lead of a Heart. Claude Rodrigue: 'How nice. That eliminates one of my possible losers. Provided I don't lose two Spade tricks, I'm going to make three Clubs, three Spades, four Diamonds and two Hearts. Now; how could I lose two Spade tricks? I can make a safety play that ensures if there is Queen to four in either hand I can catch it for the loss of one trick. I'll play a Spade to the King and if everybody follows, I will play a Spade to the A 9 7 and cover whatever South plays.'

Many players would lead the ♣A, intending to finesse the ♣J on the next round. But this is a poor play that loses two tricks unnecessarily whenever there is a small singleton in the North hand.

Board Thirty-two

Dealer South   Love All

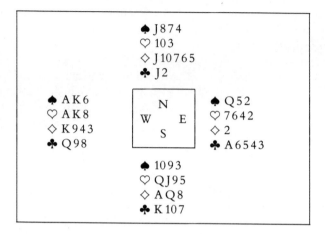

		♠ J874		
		♡ 103		
		♢ J10765		
		♣ J2		

	♠ AK6		♠ Q52
	♡ AK8	N	♡ 7642
	♢ K943	W E	♢ 2
	♣ Q98	S	♣ A6543

		♠ 1093		
		♡ QJ95		
		♢ AQ8		
		♣ K107		

Room One

South	West	North	East
DAVIES, UK	MOSS, USA	GARDENER, UK	MITCHELL, USA
1NT[1]	Dble[2]	2♣[3]	Dble[4]
NB[5]	NB[6]	2♢[7]	NB
NB	Dble	All pass	

Contract: 2♢ Doubled by North. Lead: ♢2.

An entertaining little hand that will be cited in evidence by the adherents the weak No Trump in vindication of their methods.

1 DAVIES: 'Minimum for a One No Trump but I've got good intermediate cards, so, non-vulnerable, it's an automatic opening bid.'

2 For penalties.

3 GARDENER: 'Pat wouldn't be very pleased with this hand as dummy in 1NT Doubled so the one thing I am not going to do is pass. I could bid 2♢ trying to escape, on the other hand it might play better in Spades because Pat may have four of them but have only a doubleton Diamond. Luckily we pay the 'Rimington Wriggle' here, so I shall start escape manoeuvres by calling 2♣. It's a lot of thinking for a three count, but every undertrick after the first is going to cost 200 because presumably we're going to get doubled.'

4 MITCHELL: 'Gail and I planned that if she doubles No Trumps and the opponents take it out, then another double just shows cards. I've got an Ace, I've got some high cards, and I've got some Clubs.'

5 DAVIES: 'Nicola could have Clubs here or she could be on the way to showing me a different type of hand. Either way, it's not up to me to take action. If she has Clubs, then 2♣ is going to be our best spot. And if she hasn't, she won't stand the double.'

6 MOSS: 'Strictly speaking, that's not a penalty double; it shows a hand with some cards, perhaps six-four, and a willingness to have me bid. The chances are we'll do just as well defending 2♣ Doubled as bidding on. I'll chance it.'

7 Completing the picture of her hand. 2◇ shows both Diamonds and Spades and leaves it to partner to choose. Redouble would have indicated the RED suits and Pass would have been based upon Clubs.

Two Diamonds Doubled goes two down for a penalty of 300, but that is an excellent result for the British women provided the British men in Room Two bid and make 3NT on the E-W cards.

Room Two

South	West	North	East
GRANOVETTER, USA	RODRIGUE, UK	SILVERMAN, USA	PRIDAY, UK
NB[1]	1◇[2]	NB	1♡[3]
NB	2NT[4]	NB	3NT[5]
All pass			

Contract: 3NT by West. Lead: ♠4.

1 GRANOVETTER: 'Not a bad looking twelve with a couple of ten spots, but it is a 4–3–3–3 shape. I'd like to pick up some points. I have a choice here on our system; I can pass 12 point hands: I think I will and hope it will create a little swing.'

2 RODRIGUE: 'Lovely hand. One more Jack and I'd be opening 2NT but there's no reason not to be disciplined on this sort of hand.'

3 PRIDAY: Well, this is a pretty awful looking hand. Normally I'd bid a Heart, but on four to the Seven? The only alternative bid is 1NT, and I'm not strong enough for that, and I am too good to pass, so I'll have to bid One Heart.'

4 RODRIGUE: 'Now's the time to show my values and the balanced distribution.'

5 PRIDAY: 'I've really very slim values here. I'd like to go to game if I could; that five-card Club suit must be useful, but those wretched Hearts are so anaemic. However, I'll push on, I think.'

And despite the fact there are only six tricks on top, Rodrigue made the contract with the minimum fuss, by setting up the Club suit. With the ♣K and ◇A both well placed, he made ten tricks for a swing of 4 imps to the British.

Board Thirty-three

Dealer West   N-S Vulnerable

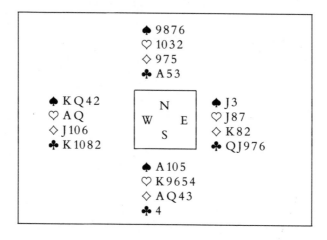

Room One

South	West	North	East
DAVIES, UK	MOSS, USA	GARDENER, UK	MITCHELL, USA
–	1NT	NB	2NT
All pass			

Contract: 2NT by West. Lead: ♠8.

Room Two

South	West	North	East
GRANOVETTER, USA	RODRIGUE, UK	SILVERMAN, USA	PRIDAY, UK
–	1♣	NB	2♣
Dble	Redble	2♠	3♣
All pass			

Contract: 3♣ by West. Lead: ♠9.

A neat example of Bridge courtesy from Tony Priday sitting East in Room Two. Rodrigue's Redouble of South's informatory Double is strength-showing and informs his partner that their side has the balance of high cards. Over Silverman's bid of 2♠, Priday realises that his hand is practically useless in defence, and so he bids 3♣ straight away instead of waiting for his partner to Double and then taking it out.

Of all the players in this match, Tony Priday is probably the one the complete palooka would prefer as a partner. Even if the palooka managed to go three off in a cast iron grand slam with 18 tricks on top, one cannot imagine Priday doing more than saying 'Oh, really!' and then going off to buy his demoralised partner a gin and tonic. There are many experts who should know better, and many lesser players who seem to delight in savaging their partners and creating an atmosphere of mutual hostility.

One really cutting sarcastic remark to partner is probably worth two tricks to the opponents on the next hand. Somehow, one cannot imagine the urbane Tony Priday has given away many tricks in this manner in his whole career.

As nine tricks were made by both declarers the Americans gained one imp.

Board Thirty-four

Dealer North   Game All

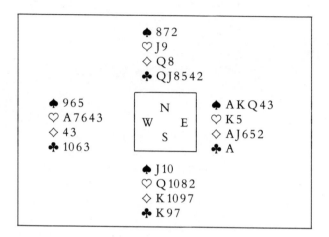

```
               ♠ 872
               ♡ J9
               ◇ Q8
               ♣ QJ8542
  ♠ 965          N          ♠ AKQ43
  ♡ A7643     W     E       ♡ K5
  ◇ 43                      ◇ AJ652
  ♣ 1063         S          ♣ A
               ♠ J10
               ♡ Q1082
               ◇ K1097
               ♣ K97
```

Room One

South	West	North	East
DAVIES, UK	MOSS, USA	GARDENER, UK	MITCHELL, USA
–	–	NB	1♠[1]
NB	NB[2]	NB	

Contract: 1♠ by East. Lead: ♡2.

This perfectly innocent-looking hand turned out to be a total horror story for the American women, and what made it worse was that they could see it coming. The East hand presents a genuine problem. All the possible opening bids, 2◇, 2♣ and 1♠, have their defects.

1 MITCHELL: 'Hey, this is a good hand. I'd like to open a two bid, but I don't have that good a suit and I don't have that many high cards nor do I have very many winners. I think I'm going to have to start the bidding with 1♠ but there's no way they are going to start with 1♠ at the other table.' (Too true, too true.)

2 MOSS: 'I hate this hand. This is a trap and I know it. The moment that Nicola and Pat both passed and I'm looking at four high-card points I know my partner has a giant hand and I'm quite sure we can make a game. But under Standard American bidding, particularly in the style that we play, I'm not allowed to respond with this hand. If I bid 2♠ in our system, there's no way I could stop my partner short of game or conceivably even a slam. It would be a partnership violation even though I know it might be the winning bid.'

Playing Standard American, the West hand, despite its four points, is too strong to pass.

Room Two

South	West	North	East
GRANOVETTER, USA	RODRIGUE, UK	SILVERMAN, USA	PRIDAY, UK
–	–	NB	2♠
NB	2NT[1]	NB	3◇[2]
NB	3♡[3]	NB	3♠[4]
NB	4♠[5]	All pass	

Contract: 4♠ by East. Lead: ♣7.

1 RODRIGUE: '2♠ could show a strong hand in Spades and possibly a second suit or it could be a weak 3♣ bid. From the opponents' silence, I think its unlikely to be the latter, but either way, I've got to make a negative bid.'

2 PRIDAY: 'Now I can show my opening was a genuine 2♠ type hand by bidding my other suit.'

3 RODRIGUE: 'Strange how a hand can suddenly improve. I've got three little trumps in support of Spades, a doubleton Diamond for ruffing values and an Ace. I think I'm even too good to bid an immediate 4♠.'

4 PRIDAY: 'Either he has a fit in one of my suits or a genuine Heart suit. I don't mind which it is; I'll just make a waiting bid and find out.'

5 RODRIGUE: 'As I've shown some interest I'll just raise to game.'

To rub salt into the wound, on the Heart lead Jackie Mitchell raked in eleven tricks while Priday just made his contract, but that was nine imps to the British and an overall lead of 45, which they held until the end of the session. The only interest on Boards 35 and 36 was the American inability to cope with the weak NT, but in the end it cost them nothing.

Board Thirty-five

Dealer East   Love All

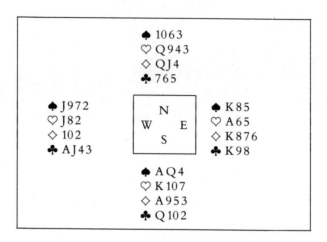

Room One

South	West	North	East
DAVIES, UK	MOSS, USA	GARDENER, UK	MITCHELL, USA
–	–	–	1◇
1NT	All pass		

Contract: 1NT by South. Lead: ♠2.

Room Two

South	West	North	East
GRANOVETTER, USA	RODRIGUE, UK	SILVERMAN, USA	PRIDAY, UK
–	–	–	1NT
Dble	NB	2♡	All pass

Contract: 2♡ by North. Lead: ◇6.

And both contracts go one down, as, probably, would have 1NT by East.

Board Thirty-six

Dealer South   Game All

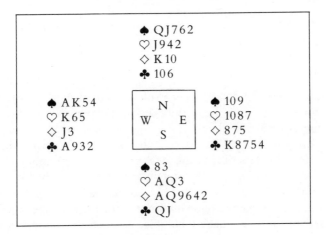

Room One

South	West	North	East
DAVIES UK	MOSS, USA	GARDENER, UK	MITCHELL, USA
1◇	Dble	1♠	NB
2◇	All pass		

Contract: 2◇ by South. Lead: ♠A.

Room Two

South	West	North	East
GRANOVETTER, USA	RODRIGUE, UK	SILVERMAN, USA	PRIDAY, UK
1◇	Dble	1♠	NB
2◇	All pass		

Contract: 2◇ by South. Lead: ◇3.

Both contracts were just made.

Session Four

In which with a hop, a skip and a jump, Uncle Sam was back on song.

Nothing pleases the English sportsman more than to see tables not merely turned, but spun. First, with the men playing against the women again, the British floundered about in the thickets of distribution on Board Thirty-nine to lose a massive 16 imps – the biggest swing in the match – then the Americans sank an approach shot for another 10 imps, only for Priday to leave a 15 foot putt six feet short on Board Forty-two to put the Americans, incredibly, in the lead.

The unfortunate Granovetter will not want to remember the last half of the session. First he made a Double of which the kindest that can be said is that it was ill-judged, and then on the last board made a mistake that cost the Americans the chance of regaining the lead they had lost to a slightly uncertain British counter-attack.

All in all, the British must have been relieved to have wobbled back into the lead by the end of the session, particularly after Board Forty-four had turned out to be a deal particularly unsuited to the aggression of the British women. Certainly in terms of the score, it was one of the more eventful sessions in the match.

Board Thirty-seven

Dealer South   N-S Vulnerable

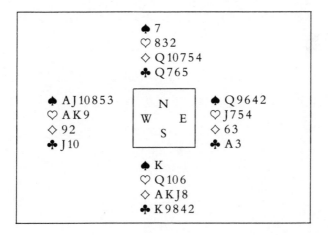

```
              ♠ 7
              ♡ 8 3 2
              ◇ Q 10 7 5 4
              ♣ Q 7 6 5

  ♠ A J 10 8 5 3      N          ♠ Q 9 6 4 2
  ♡ A K 9         W       E      ♡ J 7 5 4
  ◇ 9 2               S          ◇ 6 3
  ♣ J 10                         ♣ A 3

              ♠ K
              ♡ Q 10 6
              ◇ A K J 8
              ♣ K 9 8 4 2
```

Room One

South	West	North	East
DAVIES, UK	SILVERMAN, USA	GARDENER, UK	GRANOVETTER, USA
1♣	1♠	2♣	3♠
4♣	4♠	All pass	

Contract: 4♠ by West. Lead: ♣5.

Room Two

South	West	North	East
MITCHELL, USA	RODRIGUE, UK	MOSS, USA	PRIDAY, UK
1◇	2♠	NB	4♠
All pass			

Contract: 4♠ by West. Lead: ◇5.

In 4♠ the focal point of interest is how to tackle the Heart suit, assuming the defence attack Clubs. In Room One, a Club was led. Declarer won, drew trumps and exited with a Diamond. After cashing three tricks in the minors South led the ♡6 so Neil Silverman allowed this to run to dummy's jack.

In Room Two, Jackie Mitchell won the Diamond lead with the King and returned the ♣4, to Jack, Queen and Ace. Declarer won, drew trumps and exited with a Diamond from dummy. Jackie played her Ace and in a vain, but brave, attempt to avoid the Heart end-play exited with the ♣2 hoping her partner had the Ten and the vital ♡9. Claude Rodrigue, however, produced the ♣10 and appreciating South's dilemma, crossed to dummy with a trump to lead a Heart and finesse the Nine for eleven tricks and an imp, to Great Britain.

Board Thirty-eight

Dealer West   E-W Vulnerable

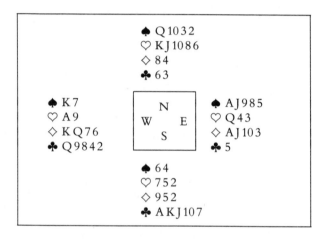

Room One

South	West	North	East
DAVIES, UK	SILVERMAN, USA	GARDENER, UK	GRANOVETTER, USA
–	1♣	1♡[1]	1♠
2♡	NB[2]	NB	Dble[3]
NB	2♠[4]	NB	3◇[5]
NB	4◇[6]	NB	5◇[7]
NB	6◇[8]	All pass	

Contract: 6◇ by East. Lead: ♣A.

1 GARDENER: 'I like to suggest a lead in this sort of position. I don't think Pat'll take it for a good hand.'

2 SILVERMAN: 'Partner's shown five Spades (a negative double would have indicated four), but I don't really want to raise him on a doubleton. I can't bid my other suit; it would show too much strength in the hand.'

3 GRANOVETTER: 'Why are these women always interfering in our auctions? Well, a Double isn't strictly for penalties.'

4 SILVERMAN: 'I like my hand, but I still have only two Spades. I think I'll just show them as secondary support.'

5 GRANOVETTER: 'I don't know: he can't have three Spades, I had to have five for my response and if he had had three, he would have raised me right away. I really don't know where this hand should be played; I'll give him my second suit and see what he has to say.'

6 SILVERMAN: 'My second suit. I'd better show some support.'

7 GRANOVETTER: '4◇? What am I going to do here. If I bid 5◇, I'm going to go down because I'll lose two Heart tricks and maybe a Club. If I bid 4♠, I might get a bad Spade break; a 5–2 fit is very precarious. Well, Neil did raise me directly to 4◇. He must have four decent Diamonds otherwise he would have put me back to 3♠. I'll just have to play 5◇.'

8 SILVERMAN: 'Let's see now. Matt's got a reasonable hand with a five-four Spade Diamond shape, just possibly even five Diamonds and five Spades. He's probably worried about the Heart control, therefore I'm almost too good to pass. Anyway, I think the likelihood is that Matt has a singleton Club, so I'm going to try 6◇.'

Room Two

South	West	North	East
MITCHELL, USA	RODRIGUE, UK	MOSS, USA	PRIDAY, UK
–	1NT[1]	NB	2♡[2]
NB	2♠[3]	NB	3◇[4]
NB	4◇[5]	NB	4♠[6]
All pass			

Contract: 4♠ by West. Lead: ♣6.

1 RODRIGUE: 'I've got the right number of points for 1NT, but I do dislike 5–4–2–2 hands for this purpose. On the other hand, if I open a suit, I can't rebid 1NT because that would show 15–17 points and if I opened 1♣, I'd have to rebid it which would put too much emphasis on the suit, and, if I bid 1◇, I'd have to rebid 2♣ (A terrible thing as Terence Reese used to say). I'm going to cheat a bit; after all, my major suits are King one and Ace one.'

2 PRIDAY: 'First duty is to show the Spade suit by means of a transfer.'

3 Automatic and a minimum or unsuitable hand.

4 PRIDAY: 'Claude doesn't seem to have any extra values and certainly not a good fit in Spades by the sound of it. I'll show him my second suit to give him the final choice for game.'

5 RODRIGUE: 'This is getting more exciting. I'm maximum; I've got a very good fit in his second suit and a moderate fit in his first so I really don't mind playing 5◇ or 6◇ if he's got a mind to it and if the worst comes to the worst and he decides to sign off in 4♠, well, King one is adequate support.'

6 Showing a lack of enthusiasm for a Slam which his partner respects.

So Rodrigue, with only a doubleton Spade, played the contract and made it for the loss of a Club, a Heart and a Spade. This was poor consolation because Granovetter had scored 1,370 for making Six Diamonds; therefore a net swing of 13 imps to the Americans. Was the rot about to set in?

Board Thirty-nine

Dealer North   Love All

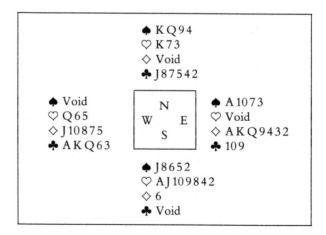

Room One

South	West	North	East
DAVIES, UK	SILVERMAN, USA	GARDENER, UK	GRANOVETTER, USA
–	–	NB[1]	1◇
4♡[2]	6◇[3]	6♡[4]	7◇[5]
NB[6]	NB[7]	7♡[8]	Dble[9]
All pass			

Contract: 7♡ Doubled by South. Lead: ♣K.

1 GARDENER: 'In my youth, I'd have opened this hand without batting an eyelid, but since then, I've become far more disciplined.'

2 DAVIES: 'On our system, I could bid 2◇ to show the major suits but if I do that, I don't think I'll ever be able to get over to Nicola how much better the Hearts are than the Spades.'

3 SILVERMAN: 'I'm just going to check the backs of these cards. There can't be all these points in the deck. I think Pat really does have a Heart suit and in that case, Matt does have a Diamond suit. I think I'll just jump to the Six level and make it tough on the opponents.'

4 GARDENER: 'With a long Heart suit, Pat can't have a lot of tricks in defence; nor have I, so I'll bid on.'

5 GRANOVETTER: 'I thought this was going to be a quiet hand and it's a head-ache. I can't believe this 6◇ bid. I have seven Diamonds to the AKQ, I mean, what is all this? He must have everything else that matters.'

6 DAVIES: 'If 7◇ is on, 7♡ is certainly going to be cheap with my distribution. In a way, I'd like to save, but I really can't do that ahead of partner. I have an Ace, she might have an Ace and it might be they've been bounced.'

7 SILVERMAN: 'I sure hope my partner wasn't expecting good trumps.'

8 GARDENER: 'Well, Pat certainly doesn't have the ♠A; she hasn't doubled, so we haven't got a certain trick. She might have a Heart trick, but against that Matthew, who is usually so slow actually bid 7◇ quickly so perhaps he thinks that he doesn't have a Heart loser. I'm going to play safe and bid one more.'

9 GRANOVETTER: 'If I pass now that would invite partner to bid 8◇. Well, not really. I guess if I pass, it invites my partner to bid 7NT. I don't have the ♡A and I remember a hand last year where the girls got to 7NT missing the Ace and went 1100 off. I'd better not do the same thing.'

Room Two

South	West	North	East
MITCHELL, USA	RODRIGUE, UK	MOSS, USA	PRIDAY, UK
–	–	NB^1	1$\diamondsuit^2$
4$\heartsuit^3$	6$\diamondsuit^4$	6$\heartsuit^5$	NB^6
NB	Dble^7	All pass	

Contract: 6$\heartsuit$ doubled by South. Lead: ♣K.

1 MOSS: 'The temptation on this hand is to pre-empt with 3♣ but Jackie has a wicked temper. Better not trifle with her.'

2 PRIDAY: 'Very awkward. I've got an Acol 2$\diamondsuit$, but I can't bid it with Claude because 2$\diamondsuit$ would show a weak Two in the majors or a 4–4–4–1 hand. Very irritating.'

3 MITCHELL: 'Well, I hope seven-five plays better than six-six.' (*A reference to the disastrous 1400 penalty on the grotesque misfit.*)

4 RODRIGUE: 'Just as I was looking forward to a nice, scientific sequence, here she comes. 4$\heartsuit$ she says. Stop, she says. I'm never going to get this hand over so I might as well take the bull by the horns and go all the way.'

5 MOSS: 'Here he is again. Every time it's my turn to bid, he's pushed me at least one level higher than the one I'd like to be at. Nothing's going to stop me supporting Jackie's Hearts, though. We might even get lucky.'

6 PRIDAY: 'Oh dear. If only we'd been playing sensible methods (*nothing can convey the anguish in his voice*) and I'd opened 2$\diamondsuit$ to start with. If I pass not only is it a forcing pass, but it announces a void in Hearts and an interest in 7$\diamondsuit$.'

7 RODRIGUE: 'In the hot seat again. Tony's pass shows that he's got first round control of Hearts and he's happy to go to 7$\diamondsuit$. Lovely as far as the Hearts are concerned, but I've leaped all the way to Six on Jack to five in support and it looks awfully foolish to bid a grand slam missing the Ace of trumps, or possibly the K and Q. I've said enough.'

On reflection, Priday rather regretted not bidding 7$\diamondsuit$ straight away, but partly because he was certain of getting a penalty and partly because he would have considered it a double-cross of partner to take out the double, he passed. The result was a disaster; the British one down in 7$\heartsuit$ doubled in Room One, the Americans making 6$\heartsuit$ doubled in Room Two resulting in the biggest swing of the match so far, 16 imps to the Americans.

Jackie Mitchell got quite a kick out of this board. She ruffed the opening lead and seeing that her only problem was to find the ♡Q she started a cross-examination of Claude Rodrigue. She could not ask Claude if he had the ♡Q but she proceeded to establish what action Tony Priday would have taken with a void in Hearts. Claude was not very forthcoming but it made no difference because at trick two Jackie led the ♡J and ran it! The ◇6 was then ruffed and trumps drawn before she tackled Spades. And with the impetus of this board, the Americans went on to grab back another 10 imps on the next.

Board Forty

Dealer East   N-S Vulnerable

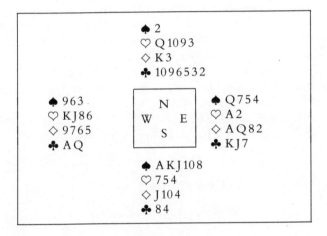

Room One

South	West	North	East
DAVIES, UK	SILVERMAN, USA	GARDENER, UK	GRANOVETTER, USA
–	–	–	1NT
NB	2♣	NB	2♦
NB	3NT	All pass	

Contract: 3NT by East. Lead: ♠K.

Room Two

South	West	North	East
MITCHELL, USA	RODRIGUE, UK	MOSS, USA	PRIDAY, UK
–	–	–	1◇
NB	1♡	NB	1♠
NB	2NT	NB	3NT
All pass			

Contract: 3NT by West. Lead: ♣5.

In Room One, Pat Davies led the ♠K showing a strong suit and requesting her partner to unblock with an honour card or signal a distributional count. If she had continued with the ♠A and the ♠J, Nicola Gardener would have had the opportunity to discard the ◇K with the object of creating an entry for South and thereby hope to defeat the contract. It probably would not have succeeded because declarer should cash the ◇AQ and North has to keep four Hearts. Clubs are played and then ♡A and another end-plays North.

In practice Pat Davies switched to the ♣8 at trick two so Matt Granovetter stripped her hand before end-playing her in Spades for a Spade return into his Queen-Seven.

In Room Two West was declarer and even after a Club lead the Spade threat was always there so Rodrigue went one down.

Board Forty-one

Dealer South   E-W Vulnerable

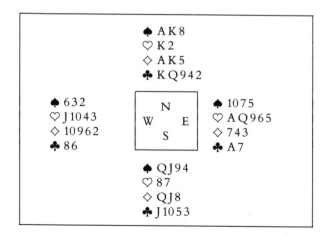

```
                    ♠ A K 8
                    ♡ K 2
                    ◇ A K 5
                    ♣ K Q 9 4 2
   ♠ 6 3 2                        ♠ 1 0 7 5
   ♡ J 1 0 4 3        N          ♡ A Q 9 6 5
   ◇ 1 0 9 6 2     W     E       ◇ 7 4 3
   ♣ 8 6              S          ♣ A 7
                    ♠ Q J 9 4
                    ♡ 8 7
                    ◇ Q J 8
                    ♣ J 1 0 5 3
```

Room One

South	West	North	East
DAVIES, UK	SILVERMAN, USA	GARDENER, UK	GRANOVETTER, USA
NB	NB	2♣	NB
2◇	NB	2NT	NB
3♣	NB	3NT	All pass

Contract: 3NT by North. Lead: ♡A.

Room Two

South	West	North	East
MITCHELL, USA	RODRIGUE, UK	MOSS, USA	PRIDAY, UK
NB	NB	2♣	NB
2◇	NB	2NT	NB
3♣	NB	3◇	NB
3NT	All pass		

Contract: 3NT by North. Lead: ♡6.

This was a very tough test of bidding. Although North-South have no less than 29 points between the two hands, North cannot make 3NT on a Heart lead (and with five to the AQ East is always going to lead a Heart) because the defence has established their suit before declarer can run the Clubs. As soon as East wins the ♣A, he can reel off four Heart tricks to defeat the contract.

The two contracts that can be made are 4♠ and 5♣, but to arrive there requires a more precise bidding system than either of the women's pairs employ.

So to Board Forty-two, the last of the half-session and a misjudgement by the British men which gave the Americans 11 imps.

Board Forty-two

Dealer West   Love All

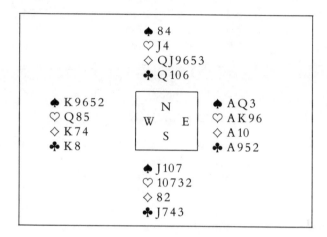

♠ 84
♡ J4
◇ QJ9653
♣ Q106

♠ K9652   ♠ AQ3
♡ Q85   ♡ AK96
◇ K74   ◇ A10
♣ K8   ♣ A952

♠ J107
♡ 10732
◇ 82
♣ J743

Room One

South	West	North	East
DAVIES, UK	SILVERMAN, USA	GARDENER, UK	GRANOVETTER, USA
–	NB	NB	2NT
NB	3♡	NB	4♠
NB	4NT	NB	5◇
NB	6♠	All pass	

Contract: 6♠ by East. Lead: ♠J.

With no interference from the British women, the American men bid carefully to reach the very good contract of 6♠. Granovetter's opening 2NT showed 20–22 points and a flat hand and his 4♠ over Silverman's transfer bid of 3♡ showed he liked the suit. After that, their system led inexorably to an unbreakable 6♠. The British men's row was made slightly harder to hoe. . .

Room Two

South	West	North	East
MITCHELL, USA	RODRIGUE, UK	MOSS, USA	PRIDAY, UK
–	NB	$2\diamond^1$	Dble^2
NB	$3\diamond^3$	NB	$3\heartsuit^4$
NB	$3\spadesuit^5$	NB	$4\spadesuit^6$
All pass			

Contract: 4♠ by West. Lead: ♦J.

1 MOSS 'Should I open this hand? It would be easily the most disgusting, deliciously low pre-emptive Two bid I could imagine. I haven't got a single first or second round control; I have no distribution; my suit is horrible. I have the minimum number of points and a partner who is disciplined to the end. Well, I've been good to her lately.'

2 PRIDAY 'The Americans seem to open weak two bids on every other hand and their values vary from practically no points to quite good hands. I must admit, it does make it very awkward. I've got a very nice 2NT opening here with no chance to bid it; but we do have our methods.' (*Sherlock Holmes' were more precise*).

3 RODRIGUE 'After passing in the first place, I'm going to have to catch up. Obviously we've got a game here, probably in Spades, but in case my partner has a maximum, I'll go the slow way. I'll cue bid the opponent's suit, then show my Spades and we'll see how it goes.'

4 PRIDAY 'Well, he's showing a few points and I suspect he's asking me to choose between the majors.'

5 Showing his useful Spade suit.

6 PRIDAY 'The more this hand goes on, the less I like it. He's showing he has Spades and he may have Clubs. 3NT is a possibility, but with ♦A 10, I don't like that. 4♣ to show the other suit is a possibility and 4♦ to show a good hand in controls is also sound. But what if Claude only has four Spades? I think I had better play this hand steadily and settle for a game which should certainly be safe.'

And for those who consider that Priday sometimes errs on the side of caution, a piece of irrefutable evidence. Twelve tricks were duly gathered at each table, but since the Americans bid the slam and the British did not, the Americans gained 11 imps.

On the half-session, a net gain of 49 imps to the USA; not only had they practically doubled their score, and wiped out the once forbidding British lead, but they had actually taken the lead for the first time since Board Fourteen.

101

Board Forty-three

Dealer West   Game All

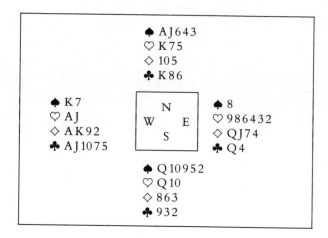

```
                    ♠ A J 6 4 3
                    ♡ K 7 5
                    ◇ 10 5
                    ♣ K 8 6

    ♠ K 7                          ♠ 8
    ♡ A J         N                ♡ 9 8 6 4 3 2
    ◇ A K 9 2   W   E              ◇ Q J 7 4
    ♣ A J 10 7 5     S             ♣ Q 4

                    ♠ Q 10 9 5 2
                    ♡ Q 10
                    ◇ 8 6 3
                    ♣ 9 3 2
```

Room One

South	West	North	East
DAVIES, UK	SILVERMAN, USA	GARDENER UK	GRANOVETTER, USA
–	1♣[1]	1♠	Dble[2]
3♠[3]	4◇[4]	All pass	

Contract: 4◇ by West. Lead: ♠A.

1 SILVERMAN 'There are two possible opening bids; I could open 2NT to show 20–22 high card points and a balanced hand or I could open 1♣ and reverse into Diamonds to show my two suits. I wouldn't like to give up the possibility of a slam; we might miss it if I bid 2NT, so I'll try the minor first.'

2 GRANOVETTER 'Gee, I don't have too much here. Not much to recommend any bid, but I do have a Heart suit so I can make a Negative Double even though I don't have more than five points. I'll risk it.'

3 DAVIES 'I've got a really bad hand here and they've obviously got something on. I'd like to make it difficult for them, by bidding 3♠ for instance, but it really is a dreadful hand and we are vulnerable. Didn't my partner call me timid a few hands back? I'm not going to let her get away with that.'

4 SILVERMAN 'I wonder if I can go back and open 2NT. Things aren't going very well here; partner's shown at least four Hearts, but I don't know; I'll have to bid my other minor.'

Although ten tricks were made, this was a wretched contract for the Americans. It all went wrong with Granovetter's double. He could not have expected his partner to bid Hearts so either he should have bid them himself straight away or, if that was forcing on their system, he should have passed and mentioned them over his partner's next bid.

Room Two

South	West	North	East
MITCHELL, USA	RODRIGUE, UK	MOSS, USA	PRIDAY, UK
–	2NT^1	NB	3$\diamondsuit^2$
NB	3$\heartsuit$	NB	4$\heartsuit$
All pass			

Contract: 4$\heartsuit$ by West. Lead: $\heartsuit$5.

1 RODRIGUE 'It seems churlish to complain when you have 20 points just because the distribution isn't exactly how you'd like it. I could open 1♣ and reverse into Diamonds if partner doesn't pass, but this is very much a hand where I want the lead coming up to me with my major suit holdings and there are enough points for 2NT.'

2 PRIDAY 'Good grief; just for once we haven't been interfered with. Well, this is fairly straight forward. Claude has 20–22 points and I have a six-card suit, anaemic though it is. I'll tell him about this with a transfer bid and let him play the hand.'

Over Rodrigue's forced response in Hearts, Priday raised to game. On the lead of the $\heartsuit$5 from North, West had to lose just the $\heartsuit$K, the ♣K and the ♠A to make the contract. The lead of the Heart deprived Rodrigue of the chance to make a simple but skilful play with only two trumps in his hand and six on the table. On any lead but a Heart, he would have crossed to dummy with a Diamond and led a small Heart to the Jack. North wins with the King, but on the next Heart lead, West goes up with the Ace and drops the Queen to lose only one trump trick. This line limits the trump loser to one when South has K10, Q10, or KQx.

Board Forty-four might have been deliberately set to trap the British women with their forthright, no nonsense aggression. They finished with the Bridge equivalent of an air-shot in golf; plenty of power but no contact.

103

Board Forty-four

Dealer North   N-S Vulnerable

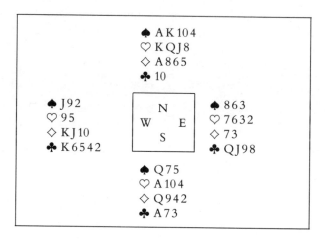

Room One

South	West	North	East
DAVIES, UK	SILVERMAN, USA	GARDENER, UK	GRANOVETTER, USA
–	–	1♡[1]	NB
2◇[2]	NB	2♠[3]	NB
3♡[4]	NB	5◇[5]	NB
6◇[6]	All pass		

Contract: 6◇ by South. Lead: ♣2.

1 GARDENER 'Usually on three-suited hands, one opens the suit below the singleton, but here the singleton is Clubs and it seems best to open 1♡ so that way we can get together in Spades or Diamonds.'

2 DAVIES 'I can't bid 2NT because in our system that shows a strong hand (16+, balanced), so for the moment I'll just bid my four-card suit and see what transpires.'

3 GARDENER 'My hand's getting better and better. Should I jump in Diamonds to agree the trump suit or show the shape of my hand first? I think I'll show her my Spades first.'

4 DAVIES 'Its beginning to sound as if Nicola has a five-card Heart suit (bidding Hearts before Spades would suggest that) so for the moment I'll just show her my Heart support and see what happens.'

5 GARDENER 'The problem about bidding my hand this way is that I've guaranteed five Hearts and now partner has given me preference. If I bid 4◇ now, she'll think I am cue-bidding the ◇A, agreeing Hearts. I suppose the only way to get over to her the fact I have four trumps with her is by jumping in Diamonds.'

6 DAVIES 'So she hasn't got five Hearts. If she had, she wouldn't have by-passed the possible 4♡ contract. So she must be 4–4–4–1 with a singleton Club' (*a fine deduction*). 'My hand looks really good now. No wasted points in Clubs with an Ace opposite her singleton; all my points must be working; I seem to have a lot of key cards including two Aces so there must be a good play for Six.'

Not with the ◇KJ10 on her left, there wasn't. It is hard to pinpoint where the bidding went off the rails. Possibly Nicola Gardener's jump to 5◇ exaggerated the quality of her Diamonds.

Room Two

South	West	North	East
MITCHELL, USA	RODRIGUE, UK	MOSS, USA	PRIDAY, UK
–	–	1◇[1]	NB
2◇[2]	NB	2♡[3]	NB
2NT[4]	NB	3NT	All pass

Contract: 3NT by South. Lead: ♣4.

1 MOSS 'At last; a nice, respectable, solid opening bid. No apologies to be made and playing five card majors, no choice to make.'

2 MITCHELL 'Well. I could bid 2NT, but this is not exactly a hand I'd prefer to play from my side of the table. So I think I'll just bid 2◇ (forcing inverted minor suit raise) and see if she'll grab the No Trumps.'

3 MOSS 'Now that's a very informative bid. It's a simple raise in Diamonds, but it tends to deny a four-card major so we're unlikely to find a four-four fit in either Hearts or Spades. But I'll bid my values so I can better judge where we are going.'

4 MITCHELL 'That's the last chance I'm giving her. I've got my Club stopper and maybe a Spade stopper.'

5 MOSS 'Back home they say that you're only supposed to play 5◇ twice a year, Christmas and birthdays. It's not Christmas and I don't think anyone is celebrating a birthday, so if Jackie wants to play No Trumps, No Trumps it is.'

105

Mitchell ducked the first two Club leads, then reeled off all the rest of the tricks, but one, for a 12 imp swing to the USA which put them back in the lead, 120 to 114.

On Board Forty-five, there seem to be chances of making 4♠, but if at the table as East you ran the ♠7 round to South's singleton ♠A, you might be accused of peeping. You probably would have been, too.

Board Forty-five

Dealer East   E-W Vulnerable

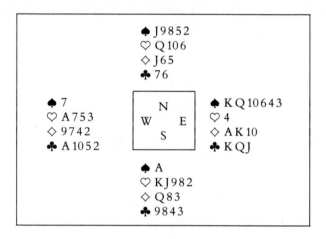

	♠ J9852	
	♡ Q106	
	◇ J65	
	♣ 76	

	♠ A	
	♡ KJ982	
	◇ Q83	
	♣ 9843	

Room One

South	West	North	East
DAVIES, UK	SILVERMAN, USA	GARDENER, UK	GRANOVETTER, USA
–	–	–	1♠
NB	1NT	NB	3♠
NB	4♠	All pass	

Contract: 4♠ by East. Lead: ♣8.

Room Two

South	West	North	East
MITCHELL, USA	RODRIGUE, UK	MOSS, USA	PRIDAY, UK
–	–	–	2♠[1]
NB	2NT	NB	3♠
NB	4♠	All pass	

Contract: 4♠ by East. Lead: ♣8.

1. A good hand with Spades or a pre-empt in Clubs.

Naturally both declarers tackled trumps by leading a top Spade from hand and then cashing the other high honour when they regained the lead. Consequently they both suffered a two-trick deficit.

Board Forty-six

Dealer South   Game All

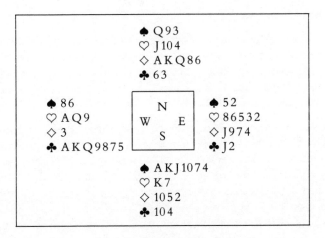

Room One

South	West	North	East
DAVIES, UK	SILVERMAN, USA	GARDENER, UK	GRANOVETTER, USA
1♠[1]	NB[2]	2◇	NB
2♠	3♣[3]	4♠[4]	NB
NB	5♣[5]	Dble	All pass

Contract: 5♣ Doubled by West. Lead: ◇A.

Room Two

South	West	North	East
MITCHELL, USA	RODRIGUE, UK	MOSS, USA	PRIDAY, UK
2♠	4♣	All pass	

Contract: 4♣ by West. Lead: ◇K.

107

1 DAVIES 'Nice shape, good suit even if minimum in points.'

2 SILVERMAN 'I could overcall 2♣ now, but I might be a little good for that. My other choice is to make a take-out double. But maybe I'll just wait a bit and see what happens.'

3 SILVERMAN 'It seems like my opponents have most of the cards, but I think I must get in there anyway; perhaps I'll get lucky and be doubled.'

4 GARDENER 'Partner's got a five-card Spade suit and she's opened the bidding. I have an opening bid, we've at least eight Spades between us; what more can we want?'

5 SILVERMAN 'The question is; can they make 4♠ or not? In all like-lihood, I only have one Club trick and maybe two Heart tricks. It looks like the Spades are breaking and I don't think we are going to be able to beat that. Looks like I've got about eight winners in my own hand; but oh well, let's go for 800.'

If the British women had held the E-W cards, this could not have happened. They have an agreement that if one of them has shown where her strength lies but she wants to bid again, then she doubles, which at least allows her partner to get into the act. And, after all, Bridge is a partnership game even if the occasional lone foray like this gives one pause to wonder.

The irony is that excellent defence can defeat 4♠. If West puts his partner in with the ♣J after cashing one high Club, then a Heart return through South wins two Heart tricks for one down. As it was, 5♣ Doubled went two down.

In Room Two, the American women quite simply got into a muddle. South's hand was too strong for an opening 2♠ under their system, but even so, with 12 points, three of partner's suit and only a doubleton of the opponent's suit, it was perhaps somewhat pusillanimous of North not to bid 4♠.

4♣ went one down undoubled; 5♣ went two down doubled, and that was 400 or 9 imps to the British to see-saw them back into the lead.

Board Forty-seven is one of those hands where it is immensely easy to be wise after the event when one has seen all 52 cards. But in truth it was a miraculous fit that might have allowed E-W to make either 6♠ or 6◇. In real life, this is the sort of contract one reaches only if pushed there by the opposition.

Board Forty-seven

Dealer West   Love All

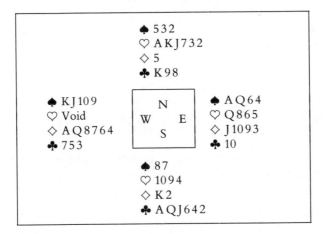

	♠ 532	
	♡ A K J 732	
	◇ 5	
	♣ K 98	
♠ K J 109		♠ A Q 64
♡ Void		♡ Q 865
◇ A Q 8764	N W E S	◇ J 1093
♣ 753		♣ 10
	♠ 87	
	♡ 1094	
	◇ K 2	
	♣ A Q J 642	

Room One

South	West	North	East
DAVIES, UK	SILVERMAN, USA	GARDENER, UK	GRANOVETTER, USA
–	1◇[1]	2♡[2]	Dble[3]
4♡[4]	4♠[5]	All pass	

Contract: 4♠ by West. Lead: ♡K.

1 SILVERMAN 'Not the strongest hand, but I do have two good suits and a void.'

2 GARDENER 'On our system, my hand qualifies for an intermediate jump overcall.'

3 GRANOVETTER 'This is quite an easy hand to bid; I can make a negative double which guarantees at least four Spades and some values.'

4 DAVIES 'On the whole, I'd expect to make 4♡ on this hand following the intermediate jump overcall. The problem is the opposition might decide to save, and it would have been useful if I'd shown Nicola that I had some points in Clubs so that she could decide what to do. But I'm not certain that 4♣ would agree a Heart fit and a Club suit so I'd better just bid 4♡.'

5 SILVERMAN 'Matthew has four Spades. Well, I certainly want to play there.'

Room Two

South	West	North	East
MITCHELL, USA	RODRIGUE, UK	MOSS, USA	PRIDAY, UK
–	1$\diamond^1$	1$\heartsuit^2$	1$\spadesuit^3$
2$\clubsuit^4$	3$\spadesuit^5$	4$\heartsuit^6$	4$\spadesuit^7$
5$\heartsuit^8$	NB	NB^9	Dble^{10}
All pass			

Contract: 5$\heartsuit$ Doubled by North. Lead: $\diamond$J.

1 RODRIGUE 'Not a lot of points, but plenty of shape. I'd open this hand every day of the week.'

2 MOSS 'I kind of wish we were playing intermediate jump overcalls, but we're not, so there's only one bid available.'

3 PRIDAY 'Good fit in Diamonds, but I must mention my four-card major first.'

4 MITCHELL 'At this table, everybody is always in the bidding. Probably, I should raise Gail's Hearts right away but the British have the higher ranking suit and I've got a lot of values in Clubs. I think I'll try to get both bids in. Bid Clubs now and then maybe support Gail's Hearts later.'

5 RODRIGUE 'Where are all the Hearts? Gail's bid 1$\heartsuit$, Tony bid a Spade instead of doubling; Jackie bids Clubs and doesn't support Hearts. My hand has improved beyond my wildest expectations when I opened the bidding. Opponents are bidding my void and Tony has bid my second suit. 3$\spadesuit$ may be a bit pushy, but I've got to get the message over.'

6 MOSS 'There he is again. But I don't think the bidding is going to die now that they've found a nice fit in Spades. My partner must be short in Spades, and since I could have overcalled with only a five-card suit but in fact have six good-looking ones, I'm going to rebid my Hearts now and when the opponents come in with 4$\spadesuit$, I'll give my partner the choice by bidding 5$\clubsuit$.'

7 PRIDAY 'Well, I could punish them by doubling, but that would be short-sighted, I think. I haven't supported my partner's Diamonds but he's given me a jump in Spades so he's certainly got four of them'.

8 MITCHELL 'I'm absolutely amazed Gail has offered her Heart suit again without me. I think I'd better support her'.

9 MOSS 'Well, even if we get doubled in Hearts, I'm sure opponents can make their game. I suspect what they probably can't know at this point is they might even have a slam.'

10 PRIDAY 'Enough is enough; I could have doubled 4♡ and I'm certainly going to double 5♡.

With the American men making an over-trick in 4♠ on the E-W cards and the women going three down doubled on the N-S cards, after the dust had settled it was 2 imps to the British.

Board Forty-eight

Dealer North   E-W Vulnerable

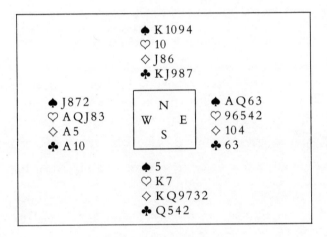

Room Two

South	West	North	East
MITCHELL, USA	RODRIGUE, UK	MOSS, USA	PRIDAY, UK
–	–	NB	NB
2◇[1]	Dble[2]	3♣[3]	3♡[4]
NB	4♡	NB	NB
5♣[5]	Dble[6]	5◇[7]	Dble
All pass			

Contract: 5◇ Doubled by South. Lead: ♡A.

1 MITCHELL 'If ever I saw a cut and dried weak two, this is it.'

2 RODRIGUE 'They don't seem to have any other bid but weak 2◇. I've got both majors and a good hand, so I'll ask Tony to choose.'

111

3 MOSS 'I've got so many options my head is swimming. Partner has a weak hand presumably with six Diamonds and it feels as if opponents probably have game in Hearts. So I could bid some large number of Diamonds to crowd them out. I think I'm going to try to be a little more scientific by bidding where my values are to make the final judgement more two-sided.'

4 PRIDAY 'I'd like to give Claude the choice of which major since he obviously has them both. I could bid 3◇ to do this, but that's rather overstating my points. Anyway, I have five Hearts.'

5 MITCHELL 'Well, I created a problem for myself by not bidding 4♣, over 3♡ and having created it, I'm going to have to solve it myself. I have no idea whether I should be defending 4♡ or sacrificing. If Gail has Clubs, we should certainly be taking a save. But if she bid it for a lead, then we're going to have a lot of chances of beating 4♡. I don't know; I just wish I'd bid 4♣, over 3♡. Oh well, here goes nothing.'

6 RODRIGUE 'Once Tony had passed and then bid only 3♡, I don't think we're likely to be able to take more than ten tricks, so I'll just have to take whatever penalties are coming my way.'

7 MOSS 'I could stand 5♣ doubled, but we must have as many Diamonds as Clubs; probably more as Jackie didn't raise the Clubs directly.'

Since Rodrigue had shown a presentable hand and Priday had an Ace, he doubled and the contract duly went two down doubled; which was worth 300 to the British. But if the American men could make 4♡ on the E-W cards in Room One, then that was going to be 620 to the USA; a net gain of 320 or eight imps. . .

Room One

South	West	North	East
DAVIES, UK	SILVERMAN, USA	GARDENER, UK	GRANOVETTER, USA
–	–	NB	NB
1◇	Dble	1♠	Dble
2◇	3◇	Dble	3♡
NB	4♡	All pass	

Contract: 4♡ by East. Lead: ◇K.

. . . And that became the final contract. For once the British women's aggression was missing. Granovetter, sitting East, ducked the opening lead of $\diamond$K, won the continuation perforce, and then disappeared into a trance. For a long time, nothing happened. Granovetter stroked his moustache, Silverman shuffled a little, Gardener and Davies gave the polite impression of having been there before. By comparison with Granovetter, Rodin's 'Thinker' was a bundle of vivacity. Silverman read the stock-market report in the *New York Times* and most of the *Wall Street Journal*. Pat Davies, not wanting to show discourtesy by actually reading at the table, almost dislocated her neck trying to read the *Financial Times* lying on the floor. Coffee came and went; time, indeed life itself, began to lack meaning.

Granovetter was bound to lose a Diamond and a Club; what he was trying to do was avoid losing a Spade and a Heart. Pat Davies had opened third in hand, not vulnerable, presumably with length in Diamonds and not strength; his partner had doubled showing the majors and Nicola Gardener might have had four Spades to justify her bid, but she also might have a singleton or void because in this situation players sometimes bluff. It was possible, of course, that she had five Spades and her partner was void, in which case for Granovetter to attempt to cross to his hand with a Spade with a view to finessing a Heart would run into a ruff. It was also possible that the $\heartsuit$K was singleton in which case the lead of the $\heartsuit$A would solve his problem.

Whereupon he called for the $\heartsuit$A! The King did not fall, he had to lose a Spade and he went one down. Somebody pointed out politely that the Club play would have been better. 'Thats what I meant to do,' he muttered, 'I pulled the wrong card, that's all.'

It was Neil Silverman, obviously fond of the mildly eccentric Granovetter in spite of being maddened by the tortoise-like pace of so much of his play, who pointed the moral of this story.

'If you thought of playing the ♣A straight away, why not do it?' he said. 'The first decision is nearly always the right one. After that, one just tries to find reasons for doing something else.'

'Ah,' Granovetter said, 'but the first decision isn't always right, is it? Not always.'

Technically, the best line of play in isolation is to cross to hand with the ♠A and finesse the $\heartsuit$Q. When this wins the $\heartsuit$A is cashed and then the ♣A followed by the ♣10. It does not matter who wins the Club or who has the Spades, the contract is certain.

As the bidding has suggested that there is a risk in this line, something must be said for cashing the ♣A, and exiting with the ♣10. The intention is to rely upon the Heart finesse and a subsequent elimination.

Board Forty-eight produced nine imps for the British instead of eight for the Americans and a British lead of 14 instead of an American lead of three: the score, 134–120.

113

Session Five

BOARDS FORTY-NINE TO SIXTY
The only moral of which is that sometimes it is wrong to be right.

On only three boards out of the twelve was there no swing; for the rest, every British thrust was instantly countered. There was what was arguably the best played hand of the match, by Jackie Mitchell of the USA; a carefully calculated piece of arithmetic by Claude Rodrigue that cost the British 14 imps and finally a little confusion between Silverman and Granovetter compounded by some unhibited kitchen Bridge bidding by Rodrigue and Priday, that cost the USA 14 imps.

What made this session so gruelling, and exciting, was that so few of the hands were simple. They varied from part scores to possible grand slams with none of those restful sequences that go 1NT–3NT with nothing more complicated to tax declarer than the choice between two finesses. Yet, quite astonishingly, at the end of Board Fifty-eight, the teams were level at 158 imps each.

Board Forty-nine

Dealer North   E-W Vulnerable

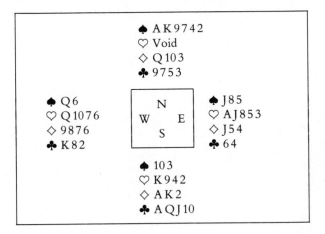

Room One

South	West	North	East
DAVIES, UK	SILVERMAN, USA	GARDENER, UK	GRANOVETTER, USA
–	–	1♠	NB
2NT[1]	NB	3♠[2]	NB
3NT	NB	4♠	All pass

Contract: 4♠ by North. Lead: ♣6.

1. 16 or more points, balanced hand (Baron, inviting partner to show a second suit.)
2. Declining to show tatty Clubs when Spades are so good.

Room Two

South	West	North	East
MITCHELL, USA	RODRIGUE, UK	MOSS, USA	PRIDAY, UK
–	–	2♠	NB
4♠	All pass		

Contract: 4♠ by North. Lead: ♣6.

1. Yet another flexible weak Two, but with a six-card suit and a void, the hand is too strong. Bidding weak Twos on hands which are as

115

strong as this can all too easily lead to good game contracts being missed.

Both contracts were made for no swing.

Board Fifty

Dealer East   Game All

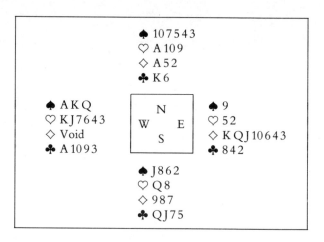

Room One

South	West	North	East
DAVIES, UK	SILVERMAN, USA	GARDENER, UK	GRANOVETTER, USA
–	–	–	3NT[1]
NB	4NT[2]	NB	5◇[3]
All pass			

Contract: 5◇ by East. Lead: ♣Q.

Room Two

South	West	North	East
MITCHELL, USA	RODRIGUE, UK	MOSS, USA	PRIDAY, UK
–	–	–	3♣
NB	3♡	NB	4◇
All pass			

Contract: 4◇ by East. Lead: ♣J.

1 GRANOVETTER 'An opening of 3◇ would show a hand not quite as good as this. We do play 3NT showing a long, nearly solid minor and I can take away three levels of bidding from the girls.'

2 SILVERMAN 'Looks like Matt over there has a long Diamond suit. We're vulnerable; I don't know, I think I'll take a shot. Maybe I'll get lucky and he has some Clubs in which case we can make a slam. I'm going to let him choose which minor and play the hand.'

3 GRANOVETTER 'Quite frankly, I don't know what that is, but I think I'll just try out my suit here.'

If, as East, you take the view the Hearts split with the Ace on your left and the Queen on your right, then you play a small one to the King, it holds and you make your contract. It is a pure guess, and the unlucky Granovetter guessed wrong; he did go up with King, but the Ace was in the North hand so he lost two Hearts and the ◇A for one down.

In Room Two, the British were playing transfer pre-empts. Priday's 3♣ bid showed Diamonds; Rodrigue took the chance to show his Hearts, and then correctly passed 4◇. Ironically, Priday guessed the Hearts right and made an over-trick!

Battered by this piece of ill fortune and still haunted by the memory of pulling the wrong card at the end of the previous session, the next board came as something of a relief to the Americans; it was the British women's turn to trip.

Board Fifty-one

Dealer South   Love All

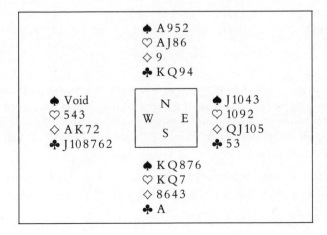

```
              ♠ A952
              ♡ AJ86
              ◇ 9
              ♣ KQ94
  ♠ Void        N        ♠ J1043
  ♡ 543     W       E    ♡ 1092
  ◇ AK72        S        ◇ QJ105
  ♣ J108762              ♣ 53
              ♠ KQ876
              ♡ KQ7
              ◇ 8643
              ♣ A
```

117

Room One

South	West	North	East
DAVIES, UK	SILVERMAN, USA	GARDENER, UK	GRANOVETTER, USA
1♠	3♣[1]	4♣[2]	NB
4♠[3]	All pass		

Contract: 4♠ by South. Lead: ◇K.

1 SILVERMAN 'Got a little suprise for them here; we play weak pre-empts at the three level.' (In many systems, a bid of 3♣ would show strength.)

2 GARDENER 'Now what am I going to do? I could bid 4♣ to show a good raise in Spades. Really, I'd like to bid 4◇ agreeing Spades and showing a shortage in Diamonds, but we don't play these complicated American systems. The best I can do is bid 4♣ and hope Pat can find some helpful bid.'

3 DAVIES 'This one hasn't come up before. Nicola could have doubled to show the other two suits. I think she must have a Spade fit and be showing a Club control. The ♣K isn't a very good card opposite my hand with its four small Diamonds. I really don't want to encourage her unless she can make a move.' But she could not.

Room Two

South	West	North	East
MITCHELL, USA	RODRIGUE, UK	MOSS, USA	PRIDAY, UK
1♠	2♣	4◇[1]	NB
4NT[2]	NB	5♡[3]	NB
6♠[4]	NB	NB	NB

Contract: 6♠ by South. Lead: ◇A.

1 MOSS 'If Claude had been silent – almost impossible to imagine – I was going to bid 4◇ (part of the complicated American system called a "splinter bid") which would show a singleton or void in Diamonds and support in Spades. Even though he's got in our way as usual, I can still make the same bid.'

2 MITCHELL 'Isn't that nice? If ever I saw 14 points that moved up, these are 14 points that moved up. All I have to worry about is Aces.'

3 Showing two Aces.

4 And so to Six.

At trick three, Jackie Mitchell made the routine expert safety play of leading a small Spade towards the Ace in dummy to guard against all four outstanding trumps being in the East hand. (If they are in the West hand, she is bound to lose a trick anyway.)

A good hand for the American bidding system and 11 imps to the United States.

Board Fifty-two

Dealer West   N-S Vulnerable

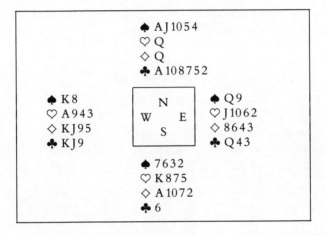

Room One

South	West	North	East
DAVIES, UK	SILVERMAN, USA	GARDENER, UK	GRANOVETTER, USA
–	1NT	2◇¹	NB
2♠	NB	3♣	NB
4♠	All pass		

Contract: 4♠ by South. Lead: ◇5.

1. Conventionally showing Spades and another suit.

Room Two

South	West	North	East
MITCHELL, USA	RODRIGUE, UK	MOSS, USA	PRIDAY, UK
–	1♡	2♣	2♡
NB	NB	2♠	NB
3♠	NB	4♠	All pass

Contract: 4♠ by North. Lead: ♡J.

In Room One, Pat Davies made two over-tricks, the lead of the ◇5 enabling her to discard dummy's ♡Q on the ◇A.

In Room Two, as Moss sitting North clearly had every intention of fighting to the end for this hand it was shrewd to bid the Clubs before the Spades. If the opponents had bid up to 4♡ Moss would have bid 4♠ showing her 6–5 distribution.

Board Fifty-three

Dealer North   Game All

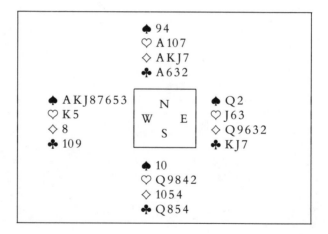

```
              ♠ 9 4
              ♡ A 10 7
              ◇ A K J 7
              ♣ A 6 3 2
♠ A K J 8 7 6 5 3      N      ♠ Q 2
♡ K 5             W    E      ♡ J 6 3
◇ 8                  S       ◇ Q 9 6 3 2
♣ 10 9                       ♣ K J 7
              ♠ 10
              ♡ Q 9 8 4 2
              ◇ 10 5 4
              ♣ Q 8 5 4
```

Room Two

South	West	North	East
MITCHELL, USA	RODRIGUE, UK	MOSS, USA	PRIDAY, UK
–	–	1NT	NB
2◇	NB[1]	2♡[2]	NB[3]
NB	3♠[4]	NB[5]	4♠[6]
All pass			

Contract: 4♠ by West. Lead: ◇K.

1 RODRIGUE 'What fun. I'm going to have a little jollity on this hand. Gail has to bid 2♡ over that 2◇, but if I were to bid 3♠ now, Tony might possibly consider it was pre-emptive even though we are vulnerable. I might lurk for a bit and see what happens if I come in later.'

2 MOSS 'My only choice is whether to bid 2♡ or to show some enthusiasm about this hand. It's not bad, but I'll just bid Two.'

3 PRIDAY (*With an air of ineffable weariness*) 'What a boring hand!'

4 RODRIGUE 'Time to come out into the open now. If I bid 3♠, Tony will know I'm not pre-empting but that I've got a goodish hand. I might almost bid 4♠; but I think I ought to let Tony bid his own hand.'

5 MOSS 'That's a surprise. I thought we were going to get to play this hand. Well, I've told my story; the rest is up to Jackie.'

6 PRIDAY 'What a turn up for the book. Well, well; Claude has a very strong suit and an excellent hand and he's inviting me to go to game if I can. My hand's pretty poor, but I do have three good cards – the ♠Q which must be valuable and the ♣KJ sitting over Gail's AQ.'

Against West's 4♠, North cashed the ◇K and led a Spade, won by declarer's King. Declarer led the ♣10, North went up with her Ace, cashed the ♡A and led the ◇A and that was the end of the story; 4♠ made by Claude Rodrigue. In Room One, things had not been going so smoothly for West after he had landed himself in the same contract.

Room One

South	West	North	East
DAVIES, UK	SILVERMAN, USA	GARDENER, UK	GRANOVETTER, USA
–	–	1◇	NB
NB	4♠	All pass	

Contract, 4♠ by West. Lead: ◇A.

121

Nicola Gardener, North, found a very neat defence. After cashing the ◇A, she led ♣A – and another Club. Her reasoning was that if West had the ♡KQ and the ♣Q he was bound to make the contract. To have his bid, he had to have either the ♡K or the ♣Q. If South had the ♣Q, then they could defeat the contract, so she assumed South did have the ♣Q and led a second Club.

This put Silverman on the spot. If he goes up with the ♣K, assuming North who opened the bidding probably has the ♡A, he must lose two Heart tricks. So he played the Jack, hoping the ♣K would provide a discard for one of his Hearts. Whereupon South won and played a Club back. West discarded one Heart, couldn't park the other, and in the end lost two Clubs, a Diamond and a Heart for one down and a well considered defence by Nicola which earned a swing of 12 imps to the UK. Nicola realised after leading the ◇A that she had too much. Unless she killed dummy's Club entry declarer would embarrass her by running his trumps, and she would succumb to a strip squeeze. Ideally a low club is better than the Ace because the defence still prevails when declarer has the ♣Q.

On Board Fifty-four, the Americans got it all back.

Board Fifty-four

Dealer East   Game All

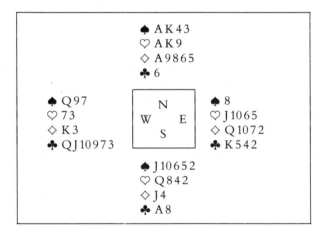

Room One

South	West	North	East
DAVIES, UK	SILVERMAN, USA	GARDENER, UK	GRANOVETTER, USA
–	–	–	NB
NB	3♣	3◇[1]	4♣[2]
4♠[3]	NB	5♣[4]	NB
5♠	All pass		

Contract: 5♠ by South. Lead: ◇3.

1 Conventional; asking partner to show her best suit.

2 GRANOVETTER 'I must say, I like my Club holding considering that partner pre-empted. But I must bear in mind he is doing it in third seat (where one often bids on even weaker hands than normal) and it is the end of the session. I'd better be a bit cautious.'

3 DAVIES 'Matthew has pinched my bid. I was going to bid 4♣ asking Nicola to pick a major. Now I'll just have to do it myself and bid my five-card suit.'

4 GARDENER 'She's never going to bid Six with all the controls I have. Still, I'm going to have a try. She could show interest by bidding 5◇.'

5 DAVIES 'So she's got a Club control and is inviting me to bid Six. But I've shown my hand and I really don't think I can do any more.'

On the lead of the ◇3, Pat Davies, because she mis-guessed the ♣Q, made exactly what she had bid – eleven tricks.

Room Two

South	West	North	East
MITCHELL, USA	RODRIGUE, UK	MOSS, USA	PRIDAY, UK
–	–	–	NB
NB	NB[1]	1◇[2]	NB
1♠[3]	NB	4♣[4]	NB
5♣[5]	NB	5◇[6]	NB
5♠[7]	NB	6♠[8]	All pass

Contract: 6♠ by South. Lead: ♡7.

1 RODRIGUE 'If one were playing weak Twos and 2♣ was not a conventional bid showing strength I could see this as an opening 2♣. But we're not, and I certainly don't see this as an opening 3♣ bid.'

2 MOSS 'Well, I have high hopes here; but I'll just start naturally by bidding my longest suit.'

3 Natural, showing at least four Spades.

4 MOSS 'Things are improving. That's my second best suit. I certainly want to play in at least a game in Spades, but I'd like to make a slam try at the same time. I can do it all with a splinter bid (showing a shortage in the bid suit, support for partner's suit and a powerful hand).'

5 MITCHELL 'Here we go again. Let's see what this hand is worth. The ♡Q is probably a good card; the Spades are good; Gail's not likely to be void in Clubs so I bet the ♣A is a working card. But what about the doubleton Diamond? I don't know whether to bid 4♠ and chicken out or move on. I think I'll make a teeny, weeny little cue bid and then I'll hold my peace forever.'

6 MOSS 'It's hard to believe my partner is strong enough to make a slam try at the five level when I'm looking at all these controls. The trump suit has to be headed by the Queen at best, she's missing the ♡K and the ◇A yet she's making a slam try. Well, I want to be in at least Six and I can even envisage Seven on this hand. I'm going to take it slowly and cue bid right back.'

7 MITCHELL 'No problems.'

8 MOSS 'Too bad. I would have loved to have heard about the ◇K. But since she doesn't seem to have that card or anything else to tell me that's relevant to the hand, I guess I'll just have to content myself with a small slam.'

That was the most splendid piece of aggression with minimum values by Jackie Mitchell, who then played most skilfully to make the slam. She won the Heart lead in hand, led a small Spade to the Ace, returned to hand with the ♣A and ran the ♠J. A third round of Spades drew the last trump. This left her with one Spade on the table to take care of the Club loser and two in her hand to establish the Diamonds after she had given up the inevitable trick in that suit. A rousing end to the half-session for the Americans to gain 13 imps and trail by 144 imps to 154; at this level, neck and neck.

Board Fifty-five

Dealer East   Love All

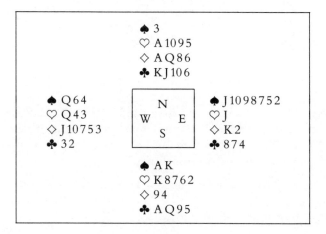

♠ 3
♡ A 10 9 5
◇ A Q 8 6
♣ K J 10 6

♠ Q 6 4
♡ Q 4 3
◇ J 10 7 5 3
♣ 3 2

N
W   E
S

♠ J 10 9 8 7 5 2
♡ J
◇ K 2
♣ 8 7 4

♠ A K
♡ K 8 7 6 2
◇ 9 4
♣ A Q 9 5

Room One

South	West	North	East
DAVIES, UK	SILVERMAN, USA	GARDENER, UK	GRANOVETTER, USA
–	–	–	3♠
3NT	NB	4♣	NB
4♡	NB	6♡	All pass

Contract: 6♡ by South. Lead: ◇J.

Despite Granovetter's pre-empt, the British women sailed serenely into 6♡. After winning the ◇J in dummy with the Ace, Pat Davies crossed to the ♡K in her hand. The appearance of East's ♡J was a give-away to an expert technician. Following the 'Principle of Restricted Choice', Pat knew that the ♡J was more likely to be a singleton than the doubleton QJ so she finessed the ♡9 to make twelve tricks, losing only one Diamond.

In Room Two, the bidding was altogether more tempestuous, and on a Spade lead, Jackie Mitchell found a well-reasoned line of play which led many to think that this was the best played single hand in the whole match.

125

Room Two

South	West	North	East
MITCHELL, USA	RODRIGUE, UK	MOSS, USA	PRIDAY, UK
–	–	–	3♡[1]
NB[2]	4♠[3]	Dble[4]	NB[5]
5♠[6]	NB	6♣[7]	NB[8]
6♡[9]	NB	NB[10]	NB

Contract: 6♡ by South. Lead: ♠4.

1 PRIDAY 'This is a pretty appalling hand, but we've had one or two rather bad results lately and I think I must get amongst them. I don't like the Spade suit headed by the J 10 9; it's a poor suit for a pre-emptive bid, but aggression must pay sometimes.'

2 MITCHELL 'Thanks a lot! I suppose I could bid 3NT with a double Spade stopper; I refuse to overcall 4♡ on that suit. Oh, I think I'll pass and maybe take some action later.'

3 RODRIGUE 'Once in a blue moon, that bid is going to show a balanced 27–29 point count. Usually it means he's got a weak 3♣ bid. What's the worst that can happen? If he's got a weak 3♣ opponents have at least a game and possibly a slam. If its the 27–29 variety, we've got a slam. I think he's more likely to have the weak variety and I'm going to bid on that assumption and make life all the more difficult for Gail who's going to have to come in at the five level.'

4 MOSS 'Just what I wanted, to have to try and guess whose hand it is at the five level. Can't be a coward. I have the right shape for a take-out double. It might go for 1100, but we've seen those numbers before.'

5 PRIDAY 'I knew it. You make a particularly weak pre-emptive bid and the roof falls in. Seems to me either I'm going to concede a large penalty or rather Claude is. That's not quite so bad.'

6 MITCHELL 'Thank you. At least that means I don't have to solve the problem about whether we should be in the bidding or not. Now the question is, how high? Six, Seven, or defending? With my high cards and Gail's double, I sure don't want to be defending. Let's just fiddle about a bit and see what happens.'

7 MOSS 'At least I guessed right whose hand it is. Partner's 5♠ is an invitation to the slam and for the moment, she's asking me to choose my best suit. Since they're all the same length, I might as well bid the first one I come to.'

8 PRIDAY 'At least we've put them to a guessing game. Let's hope they guess wrong.'

9 MITCHELL 'I wonder if Gail would realise I was trying for Seven if I bid 6♡? Well, since I have the higher ranking suit and could have bid Six a long time ago, perhaps she'll get the picture.'

10 MOSS 'My God is it our hand! Jackie now tells me she wanted to be in Six all the time and she had visions of Seven if I liked my hand. Well, I do like my hand, but I did force her to come in at the five level on 14 points, and that certainly can't be more than she is expecting me to have. No, I think I'll have to turn down her grand slam try.'

That is the mark of a good player, Moss values her hand in the context of *the whole auction.*

Mitchell won the Spade lead in hand and then cashed the ♡K. This wins against three trumps to an honour in the West hand because declarer can later finesse dummy's 10. But unlike Pat Davies, before committing herself to the Heart finesse, Jackie Mitchell does some calculation.

'Oh, sure wouldn't you know the Jack came up. Now I have to worry about a Heart loser and possibly a Diamond loser. Do I really think the Hearts are 3–1? I suppose they could be 2–2. As for the ◇K, who knows, everybody was in the bidding. Let's see. If I were to cash out some black suit winners and then take a Heart finesse and it loses, I could well be in a position to make the hand anyway with an end-play. I think that looks like the best way of playing it. There are seven Spades on my right (*he opened 3♡ showing long Spades*). If I lose to the ♡Q, that would be two Hearts. If he has two Clubs and two Diamonds, he'll either have to lead into my Diamond tenace on the table or, if I've eliminated his Clubs, give me a Diamond discard on his Spade. If he has three Clubs and exits with a Club, then I can play safe and take the Diamond finesse, because he will be marked with a singleton in the suit and all I shall be losing to is the ◇K. That looks right.'

So, at trick three, she cashed the ♠K, throwing a diamond from dummy. She then took two rounds of Clubs, finishing in her hand, ran the seven of Hearts and eventually lost only the ◇K. Well played indeed!

Board Fifty-six was a part-score hand a small swing to the British.

127

Board Fifty-six

Dealer South   N-S Vulnerable

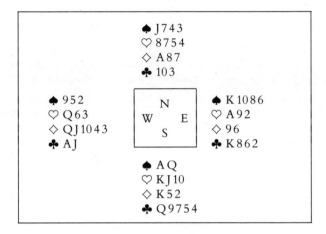

Room One

South	West	North	East
DAVIES, UK	SILVERMAN, USA	GARDENER, UK	GRANOVETTER, USA
1♣	2♢[1]	All pass	

Contract: 2♢ by West. Lead: ♣10.

1. Shows the extraordinary diversity of hands on which the Americans are prepared to make either weak openings or over-calls at the two level. Not many British players would dream of this bid on this hand, but it worked out well, Silverman coming to eight tricks.

Room Two

South	West	North	East
MITCHELL, USA	RODRIGUE, UK	MOSS, USA	PRIDAY, UK
1NT	All pass		

Contract: 1NT by South. Lead: ♢Q.

In Room Two, the British defenders also made eight tricks against South's 1NT to set it two tricks. Swing to the British: 3 imps.

Board Fifty-seven was a decidedly trappy little number and both North-Souths showed their expertise in avoiding deep and murky waters.

Board Fifty-seven

Dealer West   E-W Vulnerable

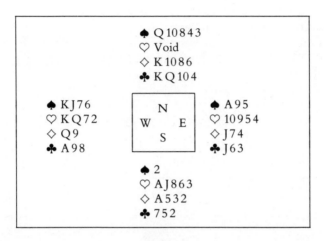

Room One

South	West	North	East
DAVIES, UK	SILVERMAN, USA	GARDENER, UK	GRANOVETTER, USA
–	1NT	2◇[1]	NB
2♡[2]	NB	2♠[3]	NB
2NT[4]	NB	3♣[5]	All pass

Contract: 3♣ by North. Lead: ◇4.

1. Astro, showing Spades and another suit.
2. Relay bid, seeking further information.
3. Five Spades. . . .
4. Asking for other suit.
5. Showing a four-card Club suit.

Room Two

South	West	North	East
MITCHELL, USA	RODRIGUE, UK	MOSS, USA	PRIDAY, UK
–	1♡	1♠	NB
1NT	NB	2♣	All pass

Contract: 2♣ by North. Lead: ♡10.

In Room One, Nicola Gardener was allowed to make an over-trick which led to an altercation between Silverman and Granovetter because without the unfortunate Diamond lead declarer should still be defeated. In Room Two, the Americans were held to eight tricks, thus giving 1 imp to the British.

Board Fifty-eight

Dealer North   Love All

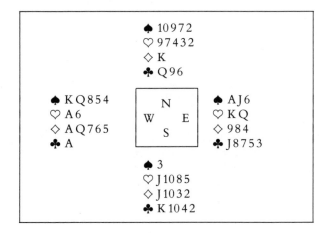

Room One

South	West	North	East
DAVIES, UK	SILVERMAN, USA	GARDENER, UK	GRANOVETTER, USA
–	–	NB	NB
NB	1♠[1]	NB	2♣[2]
NB	3◇[3]	NB	3♠[4]
NB	4NT[5]	NB	5◇[6]
NB	5NT[7]	NB	6♡[8]
NB	6♠[9]	All pass	

Contract: 6♠ by West. Lead: ♣6.

1 SILVERMAN 'What a lucky day that no-one's come in to get in my way. I've got two biddable suits so I'll start by bidding the higher ranking of them.'

2 GRANOVETTER 'Normally this could be rather an awkward hand because I wouldn't know whether to raise Spades or bid my Clubs. However, we play Drury which is a convention which shows specifically three or more trumps and ten or more points.'

3 SILVERMAN 'On that bid, we could have a slam here, even a grand slam. I may as well show him my very good second suit now.'

4 GRANOVETTER 'It seems Neil has a slam coming here in Spades or Diamonds. I don't like those three small Diamonds. I hope I can slow him down.'

5 SILVERMAN 'Now to find out how many controls Matthew has. From my point of view, I don't see how we can miss a slam now. I'd be unlucky to find him with a lot of Club and Heart cards. We'll see.'

6 GRANOVETTER 'Great. Just as I want to slow him down, he launches into Blackwood (demanding Aces).'

7 SILVERMAN 'Let's see. We have all the Aces, now I can ask him specifically which Kings he has. If he has the ♡K and the ◇K I can almost lay the cards down in 7♠.'

8 GRANOVETTER 'Well, we've all the Aces and he's strong in trumps and he's asking me to show him any extra Kings. I don't really want to because my Diamond holding is so poor he really needs ◇AKQ in his hand to make a grand slam, and I really don't want to encourage him. But if I bid 6♡, it doesn't say anything positive about my Diamonds.'

9 SILVERMAN 'Actually, that's the right King, but we don't have the ◇K otherwise he would have bid that one first.'

131

A good demonstration of key card Blackwood enables Silverman to appreciate the gap in the diamond suit.

Room Two

South	West	North	East
MITCHELL, USA	RODRIGUE, UK	MOSS, USA	PRIDAY, UK
–	–	NB	NB
NB	1♠[1]	NB	2♣[2]
NB	3◇[3]	NB	3♠[4]
NB	4♣[5]	NB	4♡[6]
NB	4NT[7]	NB	5◇[8]
NB	5♡[9]	NB	6♣[10]
NB	6♠[11]	All pass	

Contract: 6♠ by West. Lead: ♡3.

1 RODRIGUE 'A lovely hand, but a bit lacking in texture. I'll start slowly with my higher ranking suit.'

2 PRIDAY 'A chance to use one of our pet gimmicks, a Drury bid. 2♣ will show either a long, rather useless Club suit or more likely nine points or more and probably a fit in his suit.'

3 RODRIGUE 'I could bid 2◇, but that's a conventional bid, merely saying that I've got an opening bid. On the other hand, if I bid 3◇ that would get over the fact that I am five-five in the two suits and he will realise how valuable the ◇K is if he should have it.'

4 PRIDAY 'So he's got a big two-suiter; well, my first duty is to support his Spades.'

5 RODRIGUE 'Well, we've got plenty of bidding space, so I propose to use it. First, a cue bid of my lowest control.'

6 PRIDAY 'I don't like my hand much, and the three Diamonds are a disadvantage, but I must, I think, show my Heart control.'

7 RODRIGUE 'How very pleasant. Now I have no losers in Hearts and no losers in Clubs. I'm certainly not going to stop short of a small slam and I propose to investigate the grand slam. I can start off via Blackwood.'

8 Showing one Ace.

9 RODRIGUE 'Right, now we've got all the Aces. A grand slam is rather remote, particularly as I know he hasn't got the ◇K. On the other hand, if he has a singleton Diamond and enough Spades to ruff out my Diamonds, we're still going to make Seven. By bidding 6♠, I'm shutting him out, but by making another bid, I'm giving him the information that we hold all the Aces and I'm still interested in the grand slam. The bid I'm going to make theoretically asks for Kings, but he will know that isn't all I'm interested in.'

10 PRIDAY 'I find this all gets more and more depressing as the hand goes on while he gets more and more excited. I'm very tempted to deny that King and sign-off in 5♠. But I do have the ♠AJ and another and that must be of use to him. I can't bid 5NT to show the King; that would be a grand slam force, so the first available step is 6♣.'

11 RODRIGUE 'Well, he knows what I've got and if he had had no losing Diamonds he would probably have gone 5NT asking me to go to Seven with two of the top three Spade honours. As he hasn't, we'll have to settle for the small slam – confidently.'

On the lead of the ♣6, Silverman (West) reflected for a moment, then briskly gathered in twelve tricks as follows: he won the Club in hand, led a Spade to dummy's Jack, a Spade back to his Queen and a small Spade to Dummy's Ace. He re-entered his hand with the ♡A, drew the last trump with the ♠K – and led the ◇A! His reasoning was that with only one entry in dummy, the ♡K, he could finesse but once through South's hand in Diamonds. He therefore gave himself the additional chances of finding North with the King-Jack or King-Ten doubleton or the singleton ◇K.

Rodrigue reasoned differently: 'If the Diamonds are 3–2 which is going to be about two thirds of the time, I make it when the ◇K is on my right. But I can give myself extra chances by taking the deep finesse, in other words by running the Nine from dummy if it's not covered. This will work when the Jack and Ten are on my right. That ought to work quite close to 10% of the time. In addition, 50% of the time, the ◇K will be on my right. If I were to make a safety play by cashing the ◇A in case there's a singleton King on my left, then draw trumps and catch J 10 to four on my right, that would work but for one thing; it's more improbable.'

Claude went on . . . and on . . . and on . . . He convinced the producer, the presenter, and even the cameramen. Having made up his mind, he won the opening Heart in hand, took three rounds of trumps ending on the table and led the doomed ◇9. Whereupon the roof fell in. North won with the King and shot back a Heart which Rodrigue had to win in dummy. He could lead a Diamond to his AQ76 through South's J 10 3, but he could never get back onto the table to do it again. Whatever happened, he was bound to lose two more tricks – a Diamond

133

and either a promoted trump in the North hand after another Diamond lead, or a Club.

The argument about the mathematical correctness of Claude Rodrigue's analysis rumbled on for days. In the end, its impeccable correctness was admitted but as someone rather unkindly pointed out, Rodrigue won the argument, but the Americans won the imps – all 14 of them to make the score level at 158 each.

Fickle fate having fingered Claude Rodrigue on board Fifty-eight now swivelled swiftly onto the Americans, which must have left Rodrigue feeling at the end of the session that there was some justice in this world, even if not much.

Board Fifty-nine

Dealer East   N-S Vulnerable

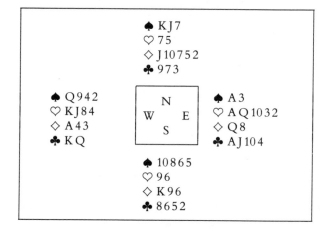

Room One

South	West	North	East
DAVIES, UK	SILVERMAN, USA	GARDENER, UK	GRANOVETTER, USA
–	–	–	1♡
NB	3♣	NB	3♡[1]
NB	3♠[2]	Dble	4♣[3]
NB	4◇[4]	NB	4♠[5]
NB	5♣[6]	NB	6♣[7]
NB	7♡[8]	All pass	

Contract: 7♡ by East. Lead: ♠5.

1 GRANOVETTER 'That's one of our conventions and I can't remember what the responses are. I think 3◇ would say I have a singleton somewhere, but all these bids; oh to be playing something natural. I don't know, I think the reply is 3♡. I sure hope so.'

2 SILVERMAN 'Well, Matt's shown two of the top three honours in Hearts. I'm going to bid 3♠ now which will show him which of the other honours I have.'

3 GRANOVETTER 'I think I've got it right. I think now that I've shown two of the top three honours and Neil has shown the other. We have a good chance of a slam here, I think. I'll make a natural call.'

4 SILVERMAN 'That's either the ♣A or a second suit. I'll bid my ◇A now.'

5 GRANOVETTER 'Very good. If Neil has the ◇AK I can get away my second Spade. I'll give him first round control in Spades.'

6 Showing the ♣K

7 GRANOVETTER 'Let's see. I've shown my Heart suit, I've shown the ♠A and my Club suit; if there's Seven on this hand, Neil would have to have the ◇AK and another so I have to hear further from him. On the other hand, there might not be 6♡; if Neil is raising Clubs now and has the KQ four times in Clubs, then its preferable to play in Clubs so that I can use my fifth Heart to pitch from Neil's hand.'

8 SILVERMAN 'He's got the ♡AQ at least five times; he's bid Clubs twice now after showing the Ace so he must be showing extra length. So he probably has five Hearts and five Clubs and the ♠A in all likelihood. So he could pitch my losing Diamonds on his Clubs; even if he has ♠A and another, he could throw Spades. Seven must be cold.'

A confidence not shared by Granovetter after Pat Davies, South, had led the ♠5: 'That's the dummy? I need another Scotch and soda. Lets see . . . five Heart tricks, four Clubs makes nine, ♠A is ten, a Diamond eleven, I can try to pitch two Diamonds from dummy then I can ruff one in my hand; that's twelve, and if I play real fast, maybe the girls will give in.'

But of course the British women did no such thing, and in the end Granovetter had to lose a Spade for one down.

Silverman must accept the blame for the impossible grand slam. Under their methods the rebid of 3♡ denies a singleton. It is thus impossible for Granovetter to hold a second five-card suit!

Room Two

South	West	North	East
MITCHELL, USA	RODRIGUE, UK	MOSS, USA	PRIDAY, UK
–	–	–	1♡
NB	4♢[1]	NB	6♡[2]
All pass			

Contract: 6♡ by East. Lead: ♠5.

1. Swiss, showing 13–15 points and a balanced hand in support of partner's opening major suit.

2. Why tell opponents anything?

Priday won the Spade lead, took two rounds of trumps and claimed the contract for the loss of just one Spade. 14 imps to the British.

The session came to a peaceful conclusion with N-S in both rooms playing in 4♠ and making Six, leaving the British leading by 172–158.

Board Sixty

Dealer South   Love All

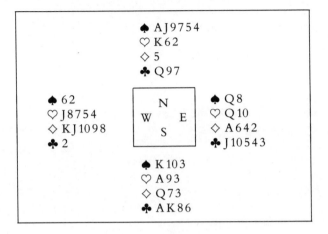

Room One

South	West	North	East
DAVIES, UK	SILVERMAN, USA	GARDENER, UK	GRANOVETTER, USA
1♣	2◇[1]	2♠	4◇
4♠	All pass		

Contract: 4♠ by North. Lead: ♣3.

1. Practically mandatory!

Room Two

South	West	North	East
MITCHELL, USA	RODRIGUE, UK	MOSS, USA	PRIDAY, UK
1NT	2◇[1]	4♠	All pass

Contract: 4♠ by North. Lead: ♡Q.

1. Showing the red suits, at least 5/4.

Session Six

'There have to be less taxing ways of making a living than playing games.'

In the last full session of twelve boards with the men playing against the men and the women against the women, eight boards were flat and only 14 imps changed hands, nine of them to the British for their lead to stretch to 18 imps.

It was as tranquil as two arm wrestlers exerting every sinew to remain motionless. The first six boards were flat, although there was a flurry on Board Sixty-three where the American men played in 3NT and the British women played the same cards in 5♣.

The deadlock was broken on Board Sixty-seven where neither team came anywhere near bidding the apparently straightforward contract of 4♠. A routine butt-in bid by Tony Priday led to such confusion in the American ranks that Silverman passed what Granovetter clearly intended to be a forcing bid. That left Granovetter to make an over-trick in the wrong part-score contract. In Room Two, Pat Davies produced another of her stunning doubles over Moss' equally normal interjection to sock the Americans for 500 and a swing of 8 imps to the British. The Americans clawed 5 imps back, so that with only six boards to go, the match could hardly have been more evenly balanced.

Board Sixty-one

Dealer West   Game All

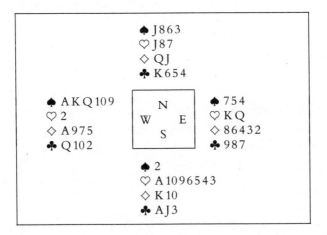

♠ J863
♡ J87
◇ QJ
♣ K654

♠ AKQ109   ♠ 754
♡ 2   ♡ KQ
◇ A975   ◇ 86432
♣ Q102   ♣ 987

♠ 2
♡ A1096543
◇ K10
♣ AJ3

Room One

South	West	North	East
RODRIGUE, UK	SILVERMAN, USA	PRIDAY, UK	GRANOVETTER, USA
–	1♠[1]	NB	1NT
3♡[2]	NB[3]	4♡[4]	All pass

Contract: 4♡ by South. Lead: ♠K.

Room Two

South	West	North	East
MITCHELL, USA	GARDENER, UK	MOSS, USA	DAVIES, UK
–	1♠	NB	2♠
4♡	4♠	5♡	All pass

Contract: 5♡ by South. Lead: ♠A.

1 Normal opening, guaranteeing five Spades.

2 RODRIGUE 'Well, I don't know what Matthew has got in mind with his forcing No Trump and my seven Hearts, but I am only interested in one thing and that is try for a Heart game.'

3 SILVERMAN 'If Claude wasn't such a nice guy, you could learn to hate him at the bridge table. Every time I have a good hand, he charges into the bidding. My Spade suit is good, but if I rebid it, it would tend to show at least six. My hand isn't that good at the four level. I hate to pass.'

4 PRIDAY 'Well, I've never understood the advantage of the forcing No Trump. However, the problem is, do I have enough to raise Claude to game? It's a pretty awful hand, but on the other hand I do have three trumps, it sounds likely that he'll be short in Spades and my ◇QJ and ♣K will be very useful.'

Silverman led the ♣K followed by the ◇A and ♠A. At trick four, Granovetter won dummy's ♡J with the Queen leaving Rodrigue to make all the rest of the tricks. He ruffed the Spade continuation, drew the last trump, led a Diamond to the table and thought for a moment. Rodrigue: 'So far I know that Neil has ♠AKQ to five and the ◇A. Matthew on my right has shown up with the ♡KQ and he's responded 1NT. Now, I must find the ♣Q. If Neil's got it, I can run off all my Hearts and Diamonds and end up with the ♠J in dummy and the King and one Club. If Neil's got the ♣Q and the winning Spade, he would be squeezed. On the other hand, if the ♣Q is on my right, I've got a simple finesse. Neil's shown his opening bid. He had full values for it. ♡KQ and nothing else seems a bit minimum for a No Trump response, so I think I'll just take the straight forward line and finesse the ♣Q, much as I like playing squeezes.'

♣K cashed on the table, ♣J finessed in hand; one down. The irony of this play emerges from Priday's comment: 'I don't understand the forcing No Trump.' Neither did Rodrigue. If East had three trumps, the ♡KQ and the ♣Q he would have raised 1♠ to 2♠. The sequence he followed is weaker.

In Room Two, Jackie Mitchell found herself playing in 5♡ but knowing that Pat Davies' first round response of 2♠ was the weakest bid she had, she did play for the squeeze and went one down as well.

To say the least, Moss' bid of 5♡ is dubious. She has four Spades to the Jack and her attacking values are too nebulous to imagine that 5♡ will succeed.

Board Sixty-two

Dealer North   N-S Vulnerable

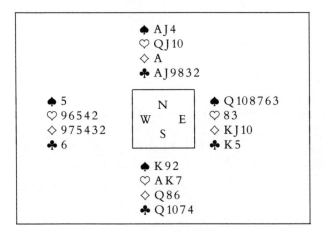

```
              ♠ A J 4
              ♡ Q J 10
              ◇ A
              ♣ A J 9 8 3 2

  ♠ 5             N            ♠ Q 10 8 7 6 3
  ♡ 9 6 5 4 2   W   E          ♡ 8 3
  ◇ 9 7 5 4 3 2     S          ◇ K J 10
  ♣ 6                          ♣ K 5

              ♠ K 9 2
              ♡ A K 7
              ◇ Q 8 6
              ♣ Q 10 7 4
```

Room One

South	West	North	East
RODRIGUE, UK	SILVERMAN, USA	PRIDAY, UK	GRANOVETTER, USA
–	–	1♣	2♠[1]
3NT[2]	NB[3]	4♣[4]	NB
4♡[5]	NB	6♣[6]	All pass

Contract: 6♣ by North. Lead: ♠8.

1 GRANOVETTER   'I think I'm getting to understand these hands now. Tony didn't open 1NT, so he could have a very good hand over there. Maybe I can cause them a little trouble.'

2 RODRIGUE   'This must be what the Americans call a pinochle deck. I've got 14 points and Tony opened the bidding and he comes in with 2♠. Well, of course it's a weak jump overcall, and I can show a Spade guard and a balanced 13–15 points.'

3 SILVERMAN   'I could get really fancy here and double. Of course they can make that, and they'll probably re-double and then when it comes back to me, I'll pull to 4◇ and cause an amazing amount of confusion. The only thing is, it might confuse my partner also. I think I'd better just wait and see what happens.'

4 PRIDAY 'Perhaps the right action would be to pass, but on the other hand I've got three controls and an excellent suit and I'm quite certain it can play satisfactorily in 5♣ if any suggestion I make is not accepted. Yes, I'm too good to pass.'

5 RODRIGUE 'That's no sign-off. Tony's looking for better things, and I'm quite delighted to co-operate.'

6 PRIDAY 'Well, that's excellent news. Claude has already shown he has a Spade stopper, almost certainly the King. Now he's shown a Heart control and implied a good fit with Clubs. I think it's a bit optimistic to hope for Seven, so I shall just settle for Six.'

And on the lead of the ♠8, he rightly rejected the Club finesse so all he had to lose was the ♣K, making twelve tricks.

Room Two

South	West	North	East
MITCHELL, USA	GARDENER, UK	MOSS, USA	DAVIES, UK
–	–	1♣	1♠
2♣[1]	NB	2◇[2]	NB
2♡[3]	NB	2♠[4]	NB
3NT[5]	NB	6♣[6]	All pass

Contract: 6♣ by North. Lead: ♡8.

1 MITCHELL 'Absolutely fantastic. I've never seen more people with more options in my life and these people bidding against me. I suppose I could bid 2NT, but this is a bad contract to play from my side of the table. I think I'll bid 2♣ (forcing, inverted minor) and see if Gail can bid 2NT from her side.'

2 MOSS 'What a pleasant bid. It's not weak, but strong, and my partner doesn't even know that I have a legitimate Club suit. I could be looking at three small Clubs. Game is a certainty, and very likely we might have a slam on this hand. I must do a bit of probing. I want to find out specifically if partner has a high card in Hearts, so I'll start off by mentioning my Diamond control.'

3 MITCHELL 'I think I'll try one more time to get her to play this silly hand in 3NT from her side.'

4 MOSS 'Splendid. Partner has at least one of the missing Heart honours so we're not off two quick tricks in Hearts if we go on to a slam in Clubs. If I bid 2♠ now, well it looks like I'm cue-bidding the ♠A but it's not really so. Partner might well think I'm looking for No Trumps and don't have the ability to bid it from my side of the table. It doesn't matter if I'm misleading her at the moment because I'll correct that impression later on. If I show Spades and bid on, it will be clear I was showing a Spade control and looking for a higher contract.'

5 MITCHELL 'I surrender.'

6 MOSS 'Ha, partner, you thought that was what I was looking for; well, you were wrong. My partner didn't know I had a Spade control, yet she was able to bid No Trumps so she must have certainly either the ♠K or Q. She's got at least one high Heart and one high Spade. Sounds to me as if she doesn't have too much wasted values in Diamonds, so a slam is a pretty good prospect. I'll stop torturing her.'

In Room One, East's lead of the ♠8 solved all declarer's problems; it gave him a free finesse in the suit, and all he had to lose was the ♣K. In Room Two, Pat Davies (East) found a much more taxing lead, the ♡8. Gail Moss: 'I can see I am potentially off a trick in both black suits. It looks as if the Spade finesse might solve the problem, but East can't have too many high card points and although it looks as if the ♠Q is in her hand, with her likely holding, I can't see the finesse working. Based on the bidding, I think East is more likely to have the ♣K as well. If she does in fact have the ♣K to go with the ♠Q and she only has one or two Clubs, but not three, then I can make this hand by getting rid of all the Diamonds in my hand and dummy, removing Pat's Hearts and then throwing her in with the ♣K. Then she'll be end-played. Either she would have to lead a Spade into my AJ or give me a ruff and discard.'

So, North won the Heart lead with the Queen in hand, cashed the ◇A, crossed to dummy with a Heart to the Ace and ruffed a Diamond. She then cashed the ♣A and crossed back to dummy with the ♠K and ruffed her last Diamond, led the ♡J to dummy's King, exited with a trump and claimed the contract.

Board Sixty-three was a mild curiosity in that the English Women went exploring for a slam and finally settled in 5♣ which they made peacefully enough, while the American men settled for a mundane 3NT. So, after three boards, no swings in this session.

Board Sixty-three

Dealer South   E-W Vulnerable

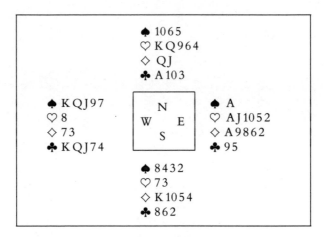

Room One

South	West	North	East
RODRIGUE, UK	SILVERMAN, USA	PRIDAY, UK	GRANOVETTER, USA
NB	1♣	1♡	NB
NB	1♠	NB	3NT
All pass			

Contract: 3NT by East. Lead: ◇4.

Room Two

South	West	North	East
MITCHELL, USA	GARDENER, UK	MOSS, USA	DAVIES, UK
NB	1♣	1♡	2◇
NB	2♠	NB	3NT
NB	4♠	NB	5♣
All pass			

Contract: 5♣ by West. Lead: ♡Q.

Board Sixty-four

Dealer South   Game All

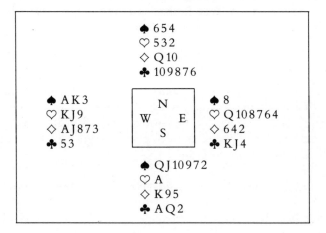

```
                ♠ 654
                ♡ 532
                ◇ Q 10
                ♣ 109876
   ♠ A K 3                    ♠ 8
   ♡ K J 9      N             ♡ Q 108764
   ◇ A J 873   W   E          ◇ 642
   ♣ 53          S            ♣ K J 4
                ♠ QJ 10972
                ♡ A
                ◇ K 95
                ♣ A Q 2
```

Room One

South	West	North	East
RODRIGUE, UK	SILVERMAN, USA	PRIDAY, UK	GRANOVETTER, USA
1♠[1]	1NT[2]	NB	2♣[3]
2♠[4]	NB[5]	NB[6]	4♡[7]
All pass			

Contract: 4♡ by East. Lead: ♣A.

1 RODRIGUE 'Its only just short of an opening Two bid. But I'll just take it slowly and open One of my six-card suit.'

2 SILVERMAN 'Yet again Claude's opened when I've got a good hand. Well, I've got 16 points and two Spade stoppers; only a couple of small Clubs, though. Well, it looks like a No Trump hand and one can't have everything.'

3 GRANOVETTER 'So partner's got 16–19 points. I've got a good chance for 4♡ if partner can help me either with maximum points or some Heart support. I think I'll start with a Stayman Two Club bid, then show him Hearts and let him decide.'

4 RODRIGUE 'Extraordinary. Here I am with a near Two bid and both opponents are bidding vulnerable. That doesn't leave much for Tony.

145

I can probably make seven tricks out of my own hand – and I can take up some of their bidding space.'

5 SILVERMAN 'Partner's asking if I have four Hearts. I don't. I don't have any surprises for Claude in the Spade suit either.' (*He means he has no telling Spade intermediate. A penalty double would be out of the question.*)

6 PRIDAY 'It's very tempting to bid 3♠ which might upset their machinery; on the other hand, I've got such an awful collection.'

7 GRANOVETTER 'Well, that's a surprise. What's Claude doing in our auction? It sounds like Claude sure has strong Spades to bid like that vulnerable. And if he has strong Spades, then Neil has weak Spades. And if Neil has weak Spades, he's got good everything else, so I'd better not give him the chance to pass.'

On the lead of the ♣A, East made ten tricks by simply ruffing a Club in dummy.

It would have been more interesting on this hand if South had led the ♠Q. Granovetter would almost certainly have played low, because he could then discard his two Diamonds on the ♠AK and establish dummy's Diamonds without allowing North to obtain the lead for a fatal Club switch. Claude's choice of the ♣A was less testing.

In Room Two, the contract was the same, but played by West instead of East.

Room Two

South	West	North	East
MITCHELL, USA	GARDENER, UK	MOSS, USA	DAVIES, UK
1♠	1NT	NB	2◇[1]
NB[2]	3♡[3]	NB	4♡[4]
NB[5]	NB	NB	

Contract: 4♡ by West. Lead: ♡2.

1 DAVIES 'Looks as if we're going to play in Hearts, but at what level? We're a bit thin on points for a game, but on the other hand, the distribution compensates for that. I'm going to bid 2◇ asking her to bid 2♡, then I can bid 3♡ showing my six-card suit and inviting her to bid game if she is suitable.'

2 MITCHELL 'I don't believe it. They're after me again. I feel like bidding 2♠. Well, Nicola's got to bid Hearts, if that passes off quietly I can come back in with 2♠.'

3 GARDENER 'I don't know whether partner is weak or strong, but my hand is eminently suitable to play in Hearts. I think I'll jump to show I have a fit and if she has a little to spare, she can give me game.'

4 DAVIES 'That solves my problems. I was going to invite game myself, so 4♡ must be a good contract.'

5 MITCHELL 'So much for bidding again.'

North led a small trump to her partner's Ace and the return of the ◇9 put Nicola on the rack straight away: 'I've lost the Ace of trumps, I probably have two Diamond losers and I'm very worried about these Clubs. It looks as if South must have the ♣AQ because she opened the bidding. But has she got the ◇KQ and would she lead the ◇9 from that holding? I could avoid a Diamond loser by pitching a Diamond on the ♠A. But I don't want Gail to get the lead. But hang on a second. Gail can't see my hand. She can't see my ♠AK or my five-card Diamond suit. And she won't believe I've let her in to lead through the ♣KJ without the Ace in my hand. If she leads anything except a Club, I'm home and dry; I think I'd better be courageous.' So West ducked the Diamond into the North hand, won the Diamond(!) return, cashed ♠AK and discarded the third Diamond from dummy, ruffed a Diamond to establish two good Diamonds in hand for Club discards, drew trumps and made 10 tricks; as audacious as the Great Train Robbery.

Board Sixty-five

Dealer East   E-W Vulnerable

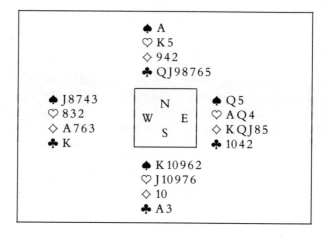

Room One

South	West	North	East
RODRIGUE, UK	SILVERMAN, USA	PRIDAY UK	GRANOVETTER, USA
–	–	–	1◇
2◇[1]	NB	2♡	NB
NB	3◇	All pass	

Contract: 3◇ by East. Lead: ♡10.

1. Michaels cue–bid indicating both major suits.

Room Two

South	West	North	East
MITCHELL, USA	GARDENER, UK	MOSS, USA	DAVIES, UK
–	–	–	1◇
2◇[1]	3◇	All pass	

Contract: 3◇ by East. Lead: ♡10.

1. Michaels cue–bid again.

A deceptive hand where over-ambition could land N-S flat on their faces, but negotiated briskly and without nonsense by experts. Notice that both North players felt it imprudent to bid their seven-card Club suit opposite South's supposed major two suiter.

Board Sixty-six presented an insoluble problem; both declarers went one down in 4♡, although the contract can be made if declarer plays the Ace of Hearts and ducks on the second round. But as one might expect both Nicola Gardener and Granovetter played with the odds, losing the ♡Q to North's King.

Board Sixty-six

Dealer West   N-S Vulnerable

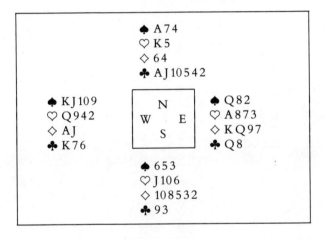

Room One

South	West	North	East
RODRIGUE, UK	SILVERMAN, USA	PRIDAY, UK	GRANOVETTER, USA
–	1♣	NB	1♡
NB	2♡	NB	3NT
NB	4♡	All pass	

Contract: 4♡ by East. Lead: ♣9.

Room Two

South	West	North	East
MITCHELL, USA	GARDENER, UK	MOSS, USA	DAVIES, UK
–	1NT	NB	2♣
NB	2♡	NB	4♡
All pass			

Contract: 4♡ by West. Lead: ◇6.

Board Sixty-seven

Dealer West   Love All

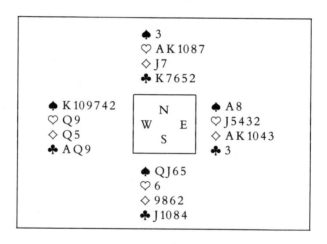

```
              ♠ 3
              ♡ A K 10 8 7
              ◇ J 7
              ♣ K 7 6 5 2
♠ K 10 9 7 4 2        ♠ A 8
♡ Q 9          N      ♡ J 5 4 3 2
◇ Q 5     W         E  ◇ A K 10 4 3
♣ A Q 9        S      ♣ 3
              ♠ Q J 6 5
              ♡ 6
              ◇ 9 8 6 2
              ♣ J 10 8 4
```

Room One

South	West	North	East
RODRIGUE, UK	SILVERMAN, USA	PRIDAY, UK	GRANOVETTER, USA
–	1♠	2♡[1]	NB[2]
NB	Dble[3]	NB[4]	3♡[5]
NB	3♠[6]	NB	4◇[7]
NB	NB[8]	NB	–[9]

Contract: 4◇ by East. Lead: ♡6.

1 PRIDAY 'Well, this is a good bad hand. I could bid 2♠ to show five Hearts and five of a minor, but it's slightly dangerous because we'd have to play at the three level and I don't think my hand quite justifies that. It does justify a bid, though, since my Hearts are quite good.'

2 GRANOVETTER 'He's taken my bid. A double is negative here, so I have to pass quite fast and not give anything away to partner; hopefully, he can re-open.' (*This is a reference to the ethical effect of a slow pass.*)

3 SILVERMAN 'Let's see; maybe my partner has a penalty double of 2♡, but in our system he's not allowed to double – that would be negative showing the other suits. I could rebid my Spades as I do have six of them, but they are not very good, so I'll treat them as five and make a re-opening double and perhaps we'll get lucky.'

4 PRIDAY 'Mmm. That's what I was afraid of. I may be going to be punished, but I think I must stick it out.'

5 GRANOVETTER 'I could pass this double for penalties and perhaps score as many as for a game contract. But if the double is for take-out, Neil could have as few as one or even no Hearts. I think I must start with a cue-bid to tell Neil that I was trapping with weak Hearts.'

6 SILVERMAN 'So on this hand he doesn't want to go for penalties. I want to try for 3NT. Maybe Matt has a side suit.'

7 GRANOVETTER 'Now I'll show him my Diamonds in case we have a fit.'

8 SILVERMAN 'Look what Speedy did to me here. Sounds like he has ◇AK to six but didn't want to bid them first time because he wasn't good enough. Hope he makes it.'

9 GRANOVETTER 'No bid? Have you gone mad; that's forcing.'

A forthright reflection on the misunderstandings which bedevil even expert partnerships. After a Heart lead to the King and a Spade return Granovetter made eleven tricks by drawing trumps and giving up a Spade to South so that he could discard his four losing Hearts on the established Spades. South, of course, did not have a second Heart to lead.

Meanwhile, things were going no more smoothly for the American women in Room Two.

Room Two

South	West	North	East
MITCHELL, USA	GARDENER, UK	MOSS, USA	DAVIES, UK
–	1♠	2♡[1]	Dble[2]
NB[3]	NB	NB[4]	

Contract: 2♡ Doubled by North. Lead: ◇A.

1 MOSS 'I wish I had a way to show both suits in one bid, but systemically, we don't. We're not vulnerable, my Hearts are quite respectable. It might be helpful to Jackie if she turns out to be on lead to let her know what I prefer.'

2 DAVIES 'I can make a penalty double here. Normally with a five-card suit, I wouldn't hesitate; the only trouble it's rather a bad five-card suit. I suppose the two things to consider are the vulnerability and whether we are likely to have a game. I can't see Gail making 2♡. But I've only 12 points and no certainty of game. It's equal vulnerability anyway; if we could make 400 from 3NT we've got a good chance of taking 300 out of 2♡ Doubled. I'll go for the penalty.'

3 MITCHELL 'The minute you hold this sort of hand, your partner is in the act. I hope they've got a slam, that's all.'

4 MOSS 'I'm not delighted with these developments. Usually you just have to take your medicine and pass. However, if Pat is so well fixed with Hearts, I might trot out my hidden suit and hope to find a happier home. I could bid 3♣, but I don't think I ought to.'

2♡ Doubled went three down to give the British 500, a gain of 8 imps and the first swing there had been for eight boards. 4♠ requires only elementary care. Declarer succeeds by playing a trump to dummy's Ace and then running the ♠8 through South.

Board Sixty-eight

Dealer North   E-W Vulnerable

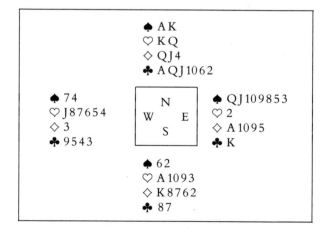

Room One

South	West	North	East
RODRIGUE, UK	SILVERMAN, USA	PRIDAY, UK	GRANOVETTER, USA
–	–	2♣	NB
3◇	NB	4♣	NB
4♡	NB	4♠	NB
5♣	All pass		

Contract: 5♣ by North. Lead: ♠Q.

Room Two

South	West	North	East
MITCHELL, USA	GARDENER, UK	MOSS, USA	DAVIES, UK
–	–	2♣	3♠
NB	NB	3NT	All pass

Contract: 3NT by North. Lead: ♠Q.

Both declarers made twelve tricks by rejecting the club finesse. But since the Americans were in No Trumps, they gained 70 points and 2 imps.

The amusement of this hand lies in what might have been. Suppose North plays 6NT after East had pre-empted in Spades. Declarer starts by developing the Diamonds; when he discovers the 4/1 break, he cashes the ♡K. East is known to have seven Spades, four Diamonds and at least one Heart. So the Clubs must be 4–1 at best. The finesse cannot therefore help, so declarer is forced to hope that East has the singleton King. Hallelujah.

Board Sixty-nine

Dealer East   Game All

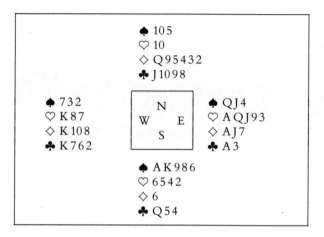

Room One

South	West	North	East
RODRIGUE, UK	SILVERMAN, USA	PRIDAY, UK	GRANOVETTER, USA
–	–	–	1♡
1♠	2♡	NB	4♡
All pass			

Contract: 4♡ by East. Lead: ♠K.

Room Two

South	West	North	East
MITCHELL, USA	GARDENER, UK	MOSS, USA	DAVIES, UK
–	–	–	2NT
NB	3♣[1]	NB	3♡[2]
NB	4♡	All pass	

Contract: 4♡ by East. Lead: ♡2.

1. Inquiry bid for five-card major suit.
2. Showing five Hearts.

After the lead of the ♠K, followed by the ♠A and a Spade ruff, the

♣J return appears to give declarer a guess. But not an expert declarer, because trumps are drawn and the ♣K cashed and a Club ruffed. This discloses that South started with five Spades, four Hearts and at least three Clubs. The ◇K is cashed therefore and the ◇J finessed with confidence.

So, with nine boards to go, there were still only 20 imps in it. On Board Seventy, the British bought the contract in both rooms – both contracts went one down to give 3 imps to the Americans.

Board Seventy

Dealer South   Love All

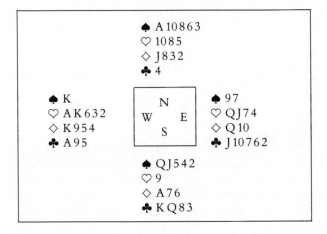

```
              ♠ A 10 8 6 3
              ♡ 1085
              ◇ J832
              ♣ 4
  ♠ K                         ♠ 97
  ♡ AK632      N              ♡ QJ74
  ◇ K954     W   E            ◇ Q10
  ♣ A95        S              ♣ J10762
              ♠ QJ542
              ♡ 9
              ◇ A76
              ♣ KQ83
```

Room One

South	West	North	East
RODRIGUE, UK	SILVERMAN, USA	PRIDAY, UK	GRANOVETTER, USA
1♠	2♡[1]	2♠	NB[2]
NB	Dble[3]	3♠[4]	4♡[5]
4♠[6]	NB[7]	NB	NB

Contract: 4♠ by South. Lead: ♡K.

1 SILVERMAN 'I'd like to make a take-out double, but when my suit is Hearts, too often we miss a 5–3 fit so I'm just going to bid my suit.'

2 GRANOVETTER 'It's very close to raise to 3♡, but I don't have an Ace, I don't have a King and I don't have a singleton.'

155

3 SILVERMAN 'Well, it couldn't have worked out better. Now I can show my other two suits.' (*This take-out double is far superior to 3◇.*)

4 PRIDAY 'Well, that's disappointing, we're not going to be allowed to play quietly in 2♠, but I was pretty good for 2♠ so certainly now I must have another little push.'

5 GRANOVETTER 'I do have four trumps for my partner which I've suppressed up to this point. I don't really fancy defending 4♠. I don't like to push them there, but on the other hand Neil re-opened with a double so he should have some values. Maybe we can beat 4♠ if they bid it, but anyway, I must show him my Heart support.'

6 RODRIGUE 'Pity I'm not being allowed to play in 2♠ or 3♠, but my defensive values really are very poor. I'm unlikely to go more than one down in 4♠; I might even have a remote chance of making it, so I'll take out a bit of insurance.'

7 SILVERMAN 'Did I say this was going well? It's become a nightmare. Opponents stopped in 2♠, now they're in Four. My partner sounds like he has four Hearts but not very many points. I don't think I'm going to beat this, but then again, I'm not going to make 5♡. I just hope we can beat them.'

The danger for the defenders here is that declarer may eliminate Hearts and Clubs, and by leading Ace and another Diamond force them to clash their Diamond honours or give him a ruff discard. After cashing the ♡K in deference to Granovetter's fine play of the ♡J, Neil Silverman switched to the ◇4 so there was never any chance of declarer succeeding.

Room Two

South	West	North	East
MITCHELL, USA	GARDENER, UK	MOSS, USA	DAVIES, UK
1♠	2♡[1]	4♣[2]	5♡[3]
All pass			

Contract: 5♡ by West. Lead: ♣4.

1 GARDENER 'My instinct tells me I should make a take-out double on this, but I've never found it works well. I'm just going to make a simple overcall. I can always double on the way back.'

2 MOSS 'If Nicola had passed, I was going to bid 4♠ which doesn't

156

show a good hand, just a hand that will play well in Spades but with very little defence.'

3 DAVIES 'Ouch! I was going to support Nicola's Hearts, but do I really want to support them at the five level? On the other hand, I don't think we're beating 4♣; I've no defence at all and Nicola's only made an overcall. Its a close thing, but let's have a go. Maybe I can push them to do the wrong thing.'

That went one off as well so the Americans had reduced the deficit to only 17 imps with eight boards to go.

Board Seventy-one was one of the most extraordinary deals in the whole match. Before you look at the bidding, cover all the cards except the East hand and decide what you would do if your partner opened a weak, pre-emptive 3♠ bid. And what do you think is the worst conceivable bid? The experts were under no illusions, but now read on.

Board Seventy-one

Dealer West   N-S Vulnerable

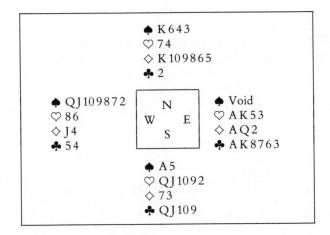

```
                    ♠ K643
                    ♡ 74
                    ◇ K109865
                    ♣ 2
    ♠ QJ109872    ┌───────────┐    ♠ Void
    ♡ 86          │    N      │    ♡ AK53
    ◇ J4          │ W     E   │    ◇ AQ2
    ♣ 54          │    S      │    ♣ AK8763
                  └───────────┘
                    ♠ A5
                    ♡ QJ1092
                    ◇ 73
                    ♣ QJ109
```

Room Two

South	West	North	East
MITCHELL, USA	GARDENER, UK	MOSS, USA	DAVIES, UK
–	3♠[1]	NB	4♠[2]
All pass			

Contract: 4♠ by West. Lead: ♣2.

1 GARDENER 'Well, I've a seven card suit; they're vulnerable and we're not, so I think I'll put a bit of pressure on Jackie and Gail here.'

2 DAVIES 'How could you do it to me, partner? One of the best hands I've had in the whole match and I have to come into the bidding at the four level. Come what may, I'm going for game. The question is, which game? No Trumps, nine tricks; 4♠, ten tricks. One thing is certain; if Nicola opens the bidding with a pre-empt, the points are going to be mainly in the suit she bid and I can't really expect her to have any tricks in No Trumps. There's no real reason to think I'm going to make nine tricks in my hand, particularly if I cannot get the Clubs going should Nicola only have a singleton or should they break badly. The only way Nicola is going to make a lot of tricks in her hand is playing the hand in Spades. Its unusual to raise partner on a void, but this seems to be the right time.'

It does not look as if there is any difficulty in making 4♠. But if South had had the ◇K it would be possible for the defence to make four tricks should he declarer unwisely play a second Club straight away. The defenders will then get a Club ruff, the two top Spades and a Diamond before declarer can establish the Clubs for a Diamond discard. So West won the ♣A, cashed the ♡AK and then ruffed a third Heart with the ♠7 – so that if North could over-ruff it would require an honour, a trick that she was going to make in any case. When North refused to over-ruff declarer led a small Club. North ruffed, and switched to the ◇6 which declarer allowed to run round to her Jack, restricting her losses to the two top Spades and a Club ruff.

Note that it doesn't help declarer to go up with the ◇A because if she does, she can never get to the table to enjoy her ♣K and so in the end must lose a Diamond anyway.

Room One

South	West	North	East
RODRIGUE, UK	SILVERMAN, USA	PRIDAY, UK	GRANOVETTER, USA
–	3♠[1]	NB	3NT[2]
NB[3]	4♠[4]	All pass	

Contract: 4♠ by West. Lead: ♣2.

1 SILVERMAN 'A long suit, not very good distribution, but the vulnerability is certainly our way so I can pre-empt the auction somewhat.'

2 GRANOVETTER 'Great. Just great, that's my void. It sort of eliminates any idea I had of bidding my hand out now. Well, I have to choose and I'm void in Spades. At this vulnerability, Neil can have a very poor Spade suit, so I'll try for nine tricks.' (*Of all the possible bids he could make, the experts unanimously agreed that this was probably the worst, and that really, in spite of the void, he had no option but to bid 4♣.*)

3 RODRIGUE 'Could they be trying to talk us out of something here? If Tony's just short of a bid over 3♣, it might be; but it's just going to be unlucky. I can't stick my neck out on this collection.'

4 SILVERMAN '3NT, partner? I don't think so. You might just have seven Clubs and two Aces, but if you do I'm just unlucky. And I can't play you for ♣AK and a small one to enter dummy.'

The play went much the same way as in Room Two, except that after ruffing the third Heart, Silverman led a trump won by South who returned a Club for partner to ruff. Back came a Diamond, allowed to run round to the Jack, which meant ten tricks.

Board Seventy-two

Dealer North   Game All

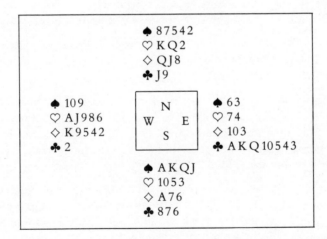

Room One

South	West	North	East
RODRIGUE, UK	SILVERMAN, USA	PRIDAY, UK	GRANOVETTER, USA
–	–	NB	NB[1]
1NT[2]	NB[3]	NB[4]	Dble[5]
NB[6]	2♡[7]	2♠[8]	3♣[9]
3♠[10]	NB[11]	NB	NB

Contract: 3♠ by North. Lead: ♡7.

1 GRANOVETTER 'Seven tricks in the Club suit, but if I pre-empt in Clubs, it shows a weaker suit in our system. I have no game or No Trump bid, so I just have to pass for now.'

2 RODRIGUE 'Fourteen points and 4–3–3–3 distribution fits our requirement for a weak No Trump, a maximum at that. I do dislike having ten of my 14 points all in one suit, but I don't want to get involved in any auction which might suggest my hand is distributional.' (*Claude's bid would certainly not please the purists.*)

3 SILVERMAN 'Two reasonable suits; I'd sure like to come in here. But, then again I have no way to show them conventionally. Some people play bids to show two suits, but unfortunately we don't. My Heart suit isn't quite good enough to come in; I just hope Matthew balances.'

4 PRIDAY 'Well, with my hand we're not going to miss game. It's possible I should bid 2♠, but with all my goodies in the other suits, I think it'll play better in 1NT.'

5 GRANOVETTER '1NT. Ha, ha! Very nice contract if my partner could lead a Club. You know, it looks like we could make 3NT if my partner had no more than an Ace and Queen over there. If I bid 3♣ now, he's still not going to know they're solid. I'd better tell him I had a maximum pass and then bid my Clubs.'

6 RODRIGUE 'Whats this maniac up to on my right? He passes originally, then when 1NT is passed round to him, he re-opens with a double. Well, I've got five tricks and hopefully Tony has something; if it's passed out, I'm looking forward to 1NT.'

7 SILVERMAN 'Very good partner. Re-opened just as I hoped he would. I think I'll show my Hearts now.'

8 PRIDAY 'The Americans obviously don't know how to handle the weak No Trump. Well, I think I should mention my Spades, dreadful though they are.'

9 GRANOVETTER '2♠. Just great. Sounds like the British need American help to reach the correct part score. What can I do except press onwards?'

10 RODRIGUE 'Curiouser and curiouser. He passes, he doubles, now he comes in at the three level. Could he have a sort of semi-solid Club suit and been lurking? I'm certainly not going to let him play in 3♣ now.'

11 SILVERMAN 'Why is it every time I look for something good to happen, it does (*the 4♠ on the previous hand*), followed by a disaster?'

West won the Heart lead with the Ace and returned a Club. East took the ♣AK and switched to the ◇10. (If he had led a third Club, West could have scored a ruff because declarer had no trump higher than the 8, and as the defence must make a Diamond the contract is defeated by one trick). The Diamond switch gave Priday the contract. He won with the Ace, drew trumps, and finished with five Spade tricks, two Hearts and two Diamonds.

Room Two

South	West	North	East
MITCHELL, USA	GARDENER, UK	MOSS, USA	DAVIES, UK
–	–	NB	3NT[1]
NB	4♣[2]	All pass	

Contract: 4♣ by West. Lead: ◇J.

1 DAVIES 'I could open 3♣ simply showing a pre-empt in Clubs, but we play an opening bid of 3NT to show a solid minor. Well, it's not completely solid – I could have the Jack – but I hate the rest of the hand, 2–2–2 and not a single point and we're vulnerable. Still, if Nicola's got quick tricks, she might forgive me not having the Jack if I've got seven Clubs.'

2 GARDENER 'Her suit must be Clubs. In No Trumps, we've got seven Club tricks, but the Spades look dodgy to say the least. I don't believe Pat has seven Clubs and the ♠A. I am going to take out some insurance here.'

From her own hand Nicola Gardener knows her partner's suit must be Clubs. 4♣ is terminal. East has to lose two Spades, a Heart and a Diamond to go one down, but the net gain to the British is 40 points and 1 imp which leaves them facing the final six boards with a lead of 18 imps, 181 to 163.

Session Seven

BOARDS SEVENTY-THREE TO SEVENTY-EIGHT
'My God, there are ghosts in this mansion.'

The final six boards began with the British leading by a mere 18 imps. The women played against the men, and after two exhausting days, all eight players knew it would take no more than one misjudgement or one stroke of brilliancy to win the match. The British had seen a lead of 45 evaporate in just six boards earlier in the match. . .

Board Seventy-three

Dealer North   Game All

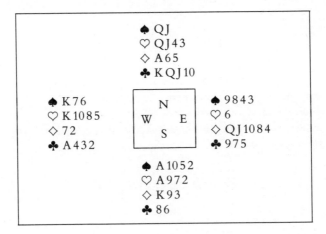

```
                    ♠ QJ
                    ♡ QJ43
                    ◇ A65
                    ♣ KQJ10
  ♠ K76          ┌─────────┐      ♠ 9843
  ♡ K1085        │    N    │      ♡ 6
  ◇ 72           │  W   E  │      ◇ QJ1084
  ♣ A432         │    S    │      ♣ 975
                 └─────────┘
                    ♠ A1052
                    ♡ A972
                    ◇ K93
                    ♣ 86
```

Room One

South	West	North	East
DAVIES, UK	SILVERMAN, USA	GARDENER, UK	GRANOVETTER, USA
–	–	1NT	NB
2♣	NB	2♡	NB
4♡	All pass		

Contract: 4♡ by North. Lead: ◇Q.

Room Two

South	West	North	East
MITCHELL, USA	RODRIGUE, UK	MOSS, USA	PRIDAY, UK
–	–	1NT	NB
2♣	NB	2♡	NB
4♡	All pass		

Contract: 4♡ by North. Lead: ◇J.

As the cards lie 4♡ should not be defeated. But in Room One they played ping-pong. The ◇Q was led which Nicola Gardener won with the King and immediately attacked Clubs. Neil Silverman played the Ace and returned a Diamond. Declarer won and cashed the ♣KQ discarding dummy's losing Diamond. Then came the ♡Q finessed to Neil

163

Silverman who held off (!) playing the ♡8, forfeiting a certain trump trick. Nicola would not rise to the bait, and switched to the ♠J which she finessed, losing to the King. Back came a Spade so she won in hand, crossed to dummy with the ♡A and played the ♣A and 10 losing but one trick to the ♡K.

If West, however, had taken his ♡K and returned a Club declarer would have lost four tricks.

In Room Two, Gail Moss was declarer and the play went similarly until trick six when she led the ♡3 to dummy's Ace and returned a low Heart to her Queen. This was better timing because when the ♠Q was finessed, West could only win and return a Spade or a Club. The ♠J was overtaken in dummy with the Ace and the ♠10 cashed, declarer discarding a Diamond. The ♡7 was led and West was helpless.

Board Seventy-four

Dealer East   Love All

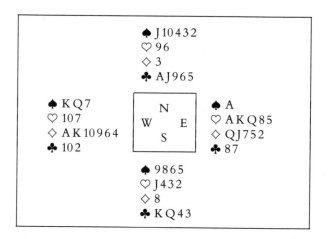

Room One

South	West	North	East
DAVIES, UK	SILVERMAN, USA	GARDENER, UK	GRANOVETTER, USA
–	–	–	1♡
NB	2♢	NB	3♢
NB	3♠	NB	4♣
Dble	NB	NB	4♢
NB	5♢	All pass	

Contract: 5♢ by West. Lead: ♣A.

Room Two

South	West	North	East
MITCHELL, USA	RODRIGUE, UK	MOSS, USA	PRIDAY, UK
–	–	–	1♡
NB	2◇	NB	4◇
NB	4♠	NB	5◇
All pass			

Contract: 5◇ by West. Lead: ♣A.

Good bidding, the same correct contract in both rooms, and no swing when the obvious eleven tricks were made. Both sides trusted the inference that there was no Club control, so they stayed safety in their depths.

Board Seventy-five

Dealer North   Love All

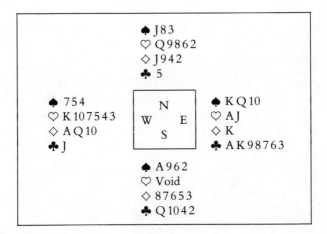

```
              ♠ J 8 3
              ♡ Q 9 8 6 2
              ◇ J 9 4 2
              ♣ 5
♠ 7 5 4              N         ♠ K Q 10
♡ K 10 7 5 4 3   W     E      ♡ A J
◇ A Q 10            S          ◇ K
♣ J                            ♣ A K 9 8 7 6 3
              ♠ A 9 6 2
              ♡ Void
              ◇ 8 7 6 5 3
              ♣ Q 10 4 2
```

Room One

South	West	North	East
DAVIES, UK	SILVERMAN, USA	GARDENER, UK	GRANOVETTER, USA
–	–	NB	2♣[1]
NB	2♠[2]	NB	3♣[3]
NB	3♡[4]	NB	3NT[5]
NB	4◇[6]	NB	4♡[7]
NB	4NT[8]	NB	NB[9]
NB			

Contract: 4NT by East. Lead: ◇5.

165

1 GRANOVETTER 'Let's see now, I have a lot of picture cards here. My God, 20 high-card points. Oh, I don't like to open a strong Two bid with only 20 points, but if I open 1♣ I really have no good rebid.'

2 SILVERMAN 'Looks like we probably have a slam here; I've just got to find out where. First I must show the Aces and Kings I have and with one Ace and one King, I can do that with a conventional bid.'

3 GRANOVETTER 'That shows an Ace and a King anywhere; but wait a second. I thought we had agreed it should be in the same suit. But look at my hand, it can't be. My God, there are ghosts in this mansion.'

4 SILVERMAN 'Better show my suit.'

5 GRANOVETTER 'I feel like I've seen this hand before, like *deja vu*. The last time I had it, I had to bid 3NT even with the singleton ◇K and I got a lead away from the Ace.'

6 SILVERMAN 'Let's see. I can bid naturally now. If he raises Diamonds. I'll take it back to No Trumps and if he raises Hearts, well let's see what happens.'

7 GRANOVETTER 'Neil's bid Hearts and Diamonds. Well, that makes my hand worth about two cents. I'll have to bid 4♡. I'd like to bid 4NT natural, but something tells me Neil will take it as Blackwood.'

8 SILVERMAN 'He may have some Heart support, or a good hand of Clubs. What to bid now? Maybe I'll bid 4NT. If he takes it as asking for Aces, I'll bid the slam. If he takes it as invitation to show a high card, I just hope he chooses the right slam.'

9 GRANOVETTER 'Neil is sitting up at the table. He looks anxious. He wants to know how many Aces I have. Well, I don't care. I feel so spooked I just want to get rid of this hand.'

Once again the Americans were at cross purposes, but on this occasion with a happier result.

Room Two

South	West	North	East
MITCHELL, USA	RODRIGUE, UK	MOSS, USA	PRIDAY, UK
–	–	NB[1]	1♣[2]
NB	1♡	NB	1♠[3]
NB	2◇[4]	NB	4♣[5]
NB	4♡[6]	NB	4NT[7]
NB	NB[8]	NB	–[9]

Contract: 4NT by East. Lead: ◇7.

1 MOSS 'Here we are getting to the last few hands and I'm dealt nothing. How are we going to pull out this match?'

2 PRIDAY 'Just my luck to have this sort of hand at this stage in the match. A very bad hand for Acol. It's a very strong hand if I open 1♣, and I might well have rebid problems. The other choices are 2♣ or 2NT. Well, it's such a bad shape even with honours everywhere, a singleton King and a seven-card suit, I think it would be ridiculous to bid 2NT. 2♣ has a lot to recommend it, except that I'm weak on points and short of controls. I think I'm stuck with the system.'

3 PRIDAY 'I might have known it. If he had bid a Diamond, I might have had an easy raise, perhaps to 3NT. Now I can't bid 3♣ which would show a strong hand but an Ace less; I can't jump in Diamonds or Spades, that would give a wrong picture of the whole hand; I can't bid 3NT with a singleton ◇K, or I wouldn't like to. I think I'll have to use a bid I use with my wife fairly frequently and bid my three-card major. It's respectable, and I'm sure it won't stay there.'

4 RODRIGUE 'Now I can bid the fourth suit. Tony can give me Heart support or preference. He may rebid a five-card Spade suit or eventually play in No Trumps. There aren't many bids that'll embarrass me.'

5 PRIDAY 'Of course, he doesn't know I've almost got a 2♣ opener. I think I must come out of the bushes and show a bit of strength. There's not a fifth suit to bid, unfortunately, so I'll have to jump in my seven-card suit.'

6 RODRIGUE 'Ha, ha! Famous last words. He's found the one bid that does embarrass me. Now where do I go? I can't pass 4♣; that would be a double-cross of the worst order. I'll just have to rebid my Heart suit and hope for the best.'

7 PRIDAY 'Now that's interesting. By this sequence, Claude has shown at least a six-card Heart suit, headed most probably by the KQ and he'll certainly have at least an Ace outside. Yes, suddenly this hand begins to look a lot more interesting. I think we should investigate a slam here. If he shows only one Ace, probably we can settle in 5♡.'

8 RODRIGUE 'Suddenly I see a haven. We might just about be able to scrape ten tricks in No Trumps. I certainly don't fancy the play for eleven tricks. I think Tony probably means this as Blackwood, but whatever he means, we're not going to go any further as far as I'm concerned.'

9 PRIDAY (Powerless further to influence events!) 'I think Claude has taken leave of his senses. Moreover, I've got to play it. Pshaw.'

Declarer won the Diamond lead in hand with his singleton King, and then played ♣AK, North discarding a small Heart on the second round. He got off lead with a small Club to South's ten, won the Diamond continuation in dummy, and reflected on his chances: 'Well, I thought this hand was going to be tricky and once the Clubs broke badly it was clear it was. Now, having discarded a Club on the second round of Diamonds which I could afford to do, I have to decide how to get to my hand to establish the Clubs. I can do one of two things. I can play a Spade to the King and hope that Gail has the Ace and ducks. After all, if she does have the Ace, she could well decide that is the right play. That would solve my entry problems and then I would have ten easy tricks. (*Four Clubs, three Diamonds, two Hearts and a Spade. It looked tempting, but the problem was that another Diamond would clear the suit and South would still have a Club entry to win her established Diamond. In fact, she would have established her Diamonds before East had established his Clubs.*) The other alternative would be to finesse the ♡J. After Gail discarded the ♡2, it does look as if she has all five of them, so that might be the successful play. I think I shall play Gail for an error, though, and lead a Spade from dummy.' But Jackie Mitchell had the ♠A and duly led a third Diamond. When she eventually won the fourth Club, she cashed a Diamond for the setting trick.

As Priday readily admits his play does not stand up to analysis. In the other room, Granovetter did finesse the ♡J and came to ten tricks, so USA gained 10 imps leaving them 8 imps behind with three boards to go.

Board Seventy-six

Dealer South   N-S Vulnerable

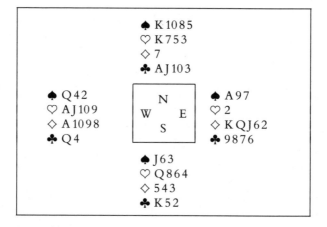

Room One

South	West	North	East
DAVIES, UK	SILVERMAN, USA	GARDENER, UK	GRANOVETTER, USA
NB	1◇	Dble	2NT[1]
NB	3◇	All pass	

Contract: 3◇ by West. Lead: ◇7.

1. Indicating a good raise in Diamonds.

Room Two

South	West	North	East
MITCHELL, USA	RODRIGUE, UK	MOSS, USA	PRIDAY, UK
NB	1NT	All pass	

Contract: 1NT by West. Lead: ♣5.

Both 3◇ and 1NT were made with an over-trick, leaving the Americans still just 8 imps behind with two boards to play.

Board Seventy-seven

Dealer West   E-W Vulnerable

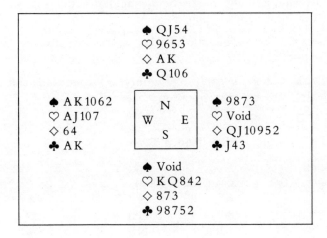

```
              ♠ Q J 5 4
              ♡ 9 6 5 3
              ◇ A K
              ♣ Q 10 6
♠ A K 10 6 2      N        ♠ 9 8 7 3
♡ A J 10 7     W     E     ♡ Void
◇ 6 4              S       ◇ Q J 10 9 5 2
♣ A K                      ♣ J 4 3
              ♠ Void
              ♡ K Q 8 4 2
              ◇ 8 7 3
              ♣ 9 8 7 5 2
```

Room One

South	West	North	East
DAVIES, UK	SILVERMAN, USA	GARDENER, UK	GRANOVETTER, USA
–	1♠	NB	4♠[1]
NB	5♣[2]	NB	5♡[3]
Dble[4]	Redble[5]	NB	5♠[6]
NB	NB[7]	NB[8]	

Contract: 5♠ by West. Lead: ♦A.

1 GRANOVETTER 'Should I bid 2♠? But that's so cowardly. I can bid 4♠ and stop any interference from Pat or Nicola. Anyway, Neil might have ♦AK, ♠AKQ and the ♣AK and then he'll go on.'

2 SILVERMAN 'Usually I would guess his second suit is going to be Diamonds, so what do I need now? Partner's telling me he's long in Spades and perhaps we just ought to play there. I'd better make just one try, though.'

3 GRANOVETTER 'Just as I thought. Partner is void in Clubs. Now I'll show him my void in Hearts.'

4 Lead directing in case they reach a slam in Spades.

5 SILVERMAN 'Is that showing a singleton or void in Hearts? I'd better show my Ace in case he has a singleton and second round control.'

6 GRANOVETTER 'So that means Neil has a high card, the ♡A or K. Which also means that I might be wrong. I'd better sign off in 5♠ quickly. I feel this hand slipping away.'

7 SILVERMAN 'I don't think this is going to work out too well; it looks like we're light two Diamond tricks, but I should be all right here.'

8 GARDENER 'There are only two contracts I'd rather defend; 6♠ or 7♠. Double? Why tell them about my Spades?'

Room Two

South	West	North	East
MITCHELL, USA	RODRIGUE, UK	MOSS, USA	PRIDAY, UK
–	1♠[1]	NB	2♠[2]
NB	4♣[3]	NB	4♡[4]
Dble	Redble[5]	NB	4♠[6]
NB	NB[7]	NB	

Contract: 4♠ by West. Lead: ♦K.

1 RODRIGUE 'Why do I never hold this sort of hand at rubber bridge? It's a near Two bid, but the suits aren't that marvellous and I'm not all that good on distribution.'

2 PRIDAY 'This looked a very poor hand, but Claude's bid has improved it a bit.'

3 RODRIGUE 'Well, I'm confident of game now; in fact, why not think of a slam? I need to know where his goodies are, particularly his controls, if any.'

4 PRIDAY 'Thats a turn-up, now he's inviting me to a slam. He's showing either a Club suit or controls and he's asking for my controls. Well, I've a pretty poor hand, but I do have a Heart void. The question is whether I am strong enough to show it. Well, I'm still seething over my play on that last hand when I went down in 4NT. I'm afraid the Americans are bound to have gained on that and are probably breathing down our necks. Once their blood is up, the American men never miss any opportunities for slams. Anyway, I don't think I will be completely ruining our cause if I tell Claude what he is asking about.'

5 RODRIGUE 'Funny, but Tony's only showing me a Heart control. I might as well let him know I have the Ace *(by Redoubling)*. It's very remote that he should have a Diamond control, but nothing will be lost if he has even if he does go to the five level to show it.'

6 PRIDAY 'I do hope he won't get too excited; he always has been very excitable, but I'm going to sign off in 4♠.'

7 RODRIGUE 'Enough is enough.'

Even more fascinating than the bidding was the play of the cards. It looks as if North must win two Diamonds and two Spades to set the contract; even when you can see all the cards. Bridge is not just a matter of technique and skill, it is a matter of exercising judgement under extreme pressure, and with a lead of only 8 imps and knowing the Americans had their tails up after the previous board, the pressure on Rodrigue was intense. But it was no less intense on Gail Moss because it was always possible that correct defence in Room Two and good play by declarer in Room One would give a swing that would put the Americans into the lead with only one board left. Gail Moss, on lead, at the first trick: 'Partner has doubled Hearts to ask me please, lead a Heart. However, I want to lay down at least one Diamond before I come to the Heart because on the bidding I'm sure my two Diamonds are likely to cash since Claude invited a slam but emphasised a weakness in Diamonds. Decision: which one to lead? Systemically, to show part-

171

ner that I have a doubleton and could ruff the third round, I should lay down the Ace first and follow it with the King. But I think this hand is one in which I want to give as little information as possible to Claude. It's going to be very tight, and I think I'd rather keep partner in the dark than elucidate Claude. So, I shall lead the ◇K.' Which held the trick, but still did not solve Gail's problem: 'Jackie has told me she doesn't want me to continue Diamonds which any fool could see looking at those Diamonds in dummy. What's the risk if I don't cash the Diamond now. Can it go away? No, there's nothing on which Claude will be able to throw away Diamonds in his hand since I've got the ♣Q. If it cashes now, it'll cash later, so I'll lead what my partner asked me to.'

On the Heart, Rodrigue discarded a small Diamond from dummy, won with the Ace in hand, led the ♣A and learned the bad news about the Spade break: 'It looks as if I have two Diamond losers and two Spade losers unless I can do something. Am I glad now that I didn't discard a club but a surplus Diamond from dummy at trick two. The ♣Q might come down so that I can discard the Diamond in my hand, but that's rather remote. I'll have to engineer a coup against Gail. If I can reduce myself by ruffing one Club in my hand and by ruffing three Hearts in dummy and being then able to exit with a Diamond, Gail will be on lead with only three cards left and will have to lead away from her ♠QJx. Provided that she follows to three Clubs and two Hearts I shall be able to do that. So I'll ruff a Heart, cash the Clubs, ruff another Heart, ruff a Club, ruff a Heart and keep following.'

And that is what happened, the luckless North being thrown in with the ◇K and having to lead away from ♠QJ5 into ♠K106. Declarer ducked the Queen but won the last two tricks. Meanwhile, in Room One, Nicola Gardener was leading ◇A and ◇K like any good kitchen Bridge player and then just sat waiting to win her inevitable Spade trick, in fact she made two trump tricks. And that swing of 13 imps on Board Seventy-seven put the match beyond the Americans' reach, giving the British a lead of 194 to 173. On the Seventy-eighth and final board, the British gained another 11 imps so that in the end they won by 205 to 173.

Board Seventy-eight

Dealer East   Love All

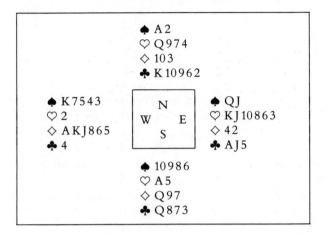

Room One

South	West	North	East
DAVIES, UK	SILVERMAN, USA	GARDENER, UK	GRANOVETTER, USA
–	–	–	2♡[1]
NB	2NT[2]	NB	3♣[3]
NB	3◇	NB	3NT
All pass			

Contract: 3NT by West. Lead: ♣10.

1. Strong weak Two bid!
2. Forcing.
3. Control, Ace or singleton.

After the ♣10 lead the defenders were in comfort and took six tricks.

Room Two

South	West	North	East
MITCHELL, USA	RODRIGUE, UK	MOSS, USA	PRIDAY, UK
–	–	–	1♡
NB	2◇	NB	2♡
NB	2♠	NB	2NT
NB	3♠	NB	4♠
All pass			

Contract: 4♠ by West. Lead: ♣6.

Despite the 4–2 Spade break, making ten tricks presented no great difficulty.

The British outbid their counterparts to gain a well-deserved swing.

Postscript

A variety of conclusions could be drawn from this match; the relative merits of British and American bidding systems; the steadiness of the American men under pressure; the element that luck plays in Bridge even at this level.

Perhaps, in the end, the difference between the two teams was the element of the unexpected. With four boards to go, the British were leading by 18 imps. 18 imps represents 1,900 points – which is exactly what Pat Davies socked the Americans for with her two penalty doubles.

Not bad for someone accused of being a timorous doubler; just remember that if you are ever in Bath and cut against her in a friendly rubber.

If you would like to know where the nearest club is to you or where Pat Davies plays Bridge so that you can put into practice all the tips you have gleaned from this match, then the English Bridge Union, 15B High Street, Thame, Oxfordshire, will be pleased to hear from you.